SCIENCE, EVOLUTION AND CREATIONISM

MODERN CONCEPTS OF NATURAL SCIENCE

SCIENCE, EVOLUTION AND CREATIONISM

Additional books in this series can be found on Nova's website under the Series tab.

Additional e-books in this series can be found on Nova's website under the e-book tab.

SCIENCE, EVOLUTION AND CREATIONISM

MODERN CONCEPTS OF NATURAL SCIENCE

MARK D. GOLDFEIN
AND
ALEXEI V. IVANOV

nova publishers
New York

For permission to use material from this book please contact us:
Telephone 631-231-7269; Fax 631-231-8175
Web Site: http://www.novapublishers.com

Additional color graphics may be available in the e-book version of this book.

Library of Congress Cataloging-in-Publication Data

ISBN: 978-1-62417-607-4

Published by Nova Science Publishers, Inc. † New York

CONTENTS

PREFACE

The monograph expounds both general and specific ideas of the formation and development of modern natural sciences, some basic problems characteristic of natural science development are discussed, integrative–general scientific trends are considered where natural-scientific knowledge is of highest value, in particular, solving global universal problems and the world community's transition to sustainable development. The book contents includes the universal laws of preservation in nature, the concept of continuous–discrete properties of matter, the foundations of the special and general theory of relativity, the laws of thermodynamics, the problem of relation of the reactivity of substances with their structure and properties, the concepts of molecular and genetic biology, the idea of the origin, structure and evolution of the Universe and Earth, the concepts of general and applied ecology, features of self-organization processes in both inanimate and animate nature.

The monograph is intended for researchers, technologists, and engineers working in various areas of natural-scientific knowledge.

INTRODUCTION

> The super-national character of scientific concepts and scientific language is due to the fact that they have been set up by the best brains of all countries and all times.
>
> *Albert Einstein*

Natural sciences as a set of scientific knowledge of natural objects, phenomena, and processes have been playing an exclusive role in the development of mankind and its culture throughout several millennia. Drawing nearer the world outlook and methodological positions of people of rather different specialties and trades is a major function of natural sciences. As the environment surrounding the man possesses a variety of levels of its organization and properties, its deep studying is only possible with the use of an interdisciplinary approach, i.e. applying the knowledge of various parts of mathematics, physics, chemistry, biology, astronomy, geology, geography, and adjacent natural sciences. While scientists, though in different degrees, have certain notions of the features of nature, the representatives of the humanities and social sciences possess, as a rule, rather limited knowledge of the natural-science picture of the world. This justifies the content of the present monograph, which includes such basic representations of the world around us as the mutual relations between the natural-scientific and humanitarian cultures (including ways of their drawing nearer), the methodology and principles of natural-scientific cognition, integration of science and education, features of the major development stages of natural sciences, the specificity of science mathematization, the major concepts of modern physics, chemistry, biology, and ecology. A special place is occupied by our discussion of some problems of cosmology, questions about the place of the man in the biosphere, the concepts of universal evolutionism and globalizations of science and engineering, the significance of the development of the informatization of the society, etc. Of doubtless interest for humanists should be our "Appendices", which include structural links among various fields of knowledge (as special schemes), data on the Nobel Prize winners in various fields of natural sciences, and a calendar of global evolutionary events.

The development history of religion, philosophy, and various sciences is related to the problem of correlation between the natural-scientific and humanitarian cultures of mankind. In recent years, in connection with new general scientific views at system analysis, self-organizing processes, and the mechanisms of the evolutionary processes in both inanimate and animate nature, the natural scientific and humanitarian cultures have begun to approach each other. In particular, in many fields of social and spiritual life, such theoretical and practical methods of natural sciences are already used as mathematical and physical modeling, new information technologies, self-organizing processes, etc. All this causes the necessity to form a wider scientific outlook and lifting the thinking culture level. The main objective of this discipline is the students' acquisition of the fullest representation about the natural scientific picture of the world. According to this purpose, the primary goals of the course are the acquisition of knowledge of most important historical stages of the development of different fields of natural sciences, the methods of natural scientific research, the general scientific-philosophical concepts, the fundamental laws of nature. All this allows drawing a fundamental conclusion of the unity of both animate and animate nature, and about the perfection of cognition of the real world, consisting in objective and regular changes of outlooks.

At the same time, the existing forms of science development cause the public opinion's concern, in increasing measure, as to the results and prospects of realization of scientific discoveries. Science has appeared responsible for slipping to a global disaster (which might occur already in the 21st century) as well as for the growth of material and spiritual goods of the modern inhabitants of our planet. This creates a real threat for economy, policy, ecology, defensibility, whole social development. The scientific potential, its ability to influence engineering, technology, education, and other spheres of social activity should be kept, of course. However, to return its lost prestige in the society, science itself should change essentially, by reorienting its development on the purposes of survival and further sustainable development of mankind. The exit from our system crisis is seen in science working within the limits of the values of the future "steady society" (the sphere of reason whose formation was predicted by Vernadsky) rather than within industrial-consumer values ruining the man and nature. It is this new-type science (noospheric-steady one) which overcoming of the arising contradictions in the man–society–biosphere system, the appearance of positive scientific preconditions for civilization survival in the 21st century and, in perspective, in the whole third millennium are associated with.

For the future specialists in the field of social sciences and humanities associated with studying the man and society, people's behavior, it is important to understand not just the essence of the concepts of modern natural sciences but also their possible influence on human activity, which will amplify in the future. That is why it is necessary to study integrative and general scientific directions of creative search, scientific knowledge synthesis during making decisions of complex (especially global) problems. The natural scientific knowledge is no longer represented as something far distant from the social humanitarian knowledge but as a component of the complete science significant for it. Until recently, the history of human life was actually disconnected from nature and started to contradict natural processes and biosphere cycles by the very fact of its existence.

The hypothesis of the eternal existence of mankind on the Earth was one of the basic positions in history and other humanities, which was very seldom discussed in science, as if mankind could transform nature with impunity and act, in relation to nature, according to the "everything for the man, for the sake of his blessing" principle. Such a hypothesis was even considered as an axiom long enough until practice began to deny its anthropocentrism by the quickly approaching global environmental crisis. This, simultaneously, was the crisis of not only the anthropo-chauvinistic" view of history but also of the idea of unlimited progress (mainly, civilization process). Science might actually appear with no subject of its research just because the approach of a planetary anthropo-ecological disaster is rather possible within few next decades, which could become omnicide, i.e. the destruction of reasonable and many other living beings. The traditional social-scientific approach has appeared principally unpromising, limited by the expansion and existence of the model of unstable development with its spatial-temporal limits. In this connection, one more feature of this book is consideration of the concepts of modern natural sciences not simply in the "socio-natural system of coordinates" but in the foreshortening of prospects of transition to sustainable development. This transition will demand the formation of such a complete socio-natural system, i.e. the mankind—nature one, which could keep its interacting components, i.e. civilization and biosphere, for vaguely long times. And from this viewpoint it is very important to emphasize those modern natural-scientific concepts which would promote the most effective movement of mankind to the sustainable development purposes.

Science, being guided on new civilization purposes, should be more and more engaged in researching the future, i.e. in perspective, the futurization of research activity and the usage of its results in practice and especially in education will proceed. Qualitative changes in science, frequent changes of paradigms and world pictures pose new problems before the researchers dealing with methodical and social-philosophical problems of education since the regularly increasing volume of scientific knowledge is difficult for laying in the accepted standards of time. There is a complex interrelation between science and education. Education depends on science because one must have knowledge before extending and broadcasting it. But knowledge is extracted by scientists. Education is therefore in direct dependence on science. On the other hand, education is primary in relation to science in the sense that the knowledge acquired in the course of education acts as a basis for new scientific results. The creation of a united advancing science–education system will demand cardinal changes in the role relations between the scientist and the teacher in the sense that the scientist should become the teacher and the teacher should become the scientist. The production of new scientific knowledge and its transfer in the course of teaching will represent, in perspective, a united scientific–educational system working for sustainable development and functioning by the principles of this civilization model.

Chapter 1

KNOWLEDGE AND SCIENCE

Those who fall in love with practice without science are like a sailor who enters a ship without a helm or a compass, and who never can be certain whither he is going. Science is the captain, and practice the soldiers.

Leonardo Da Vinci

Nothing is more practical than a good theory.

Ludwig Boltzmann

1.1. SOME NOTIONS, FEATURES AND ROLE OF NATURAL SCIENCE IN THE MODERN WORLD

Analysis of various spheres of human life and the man's interaction with the environment allows drawing a conclusion that the structure of the human world consists of four components, namely,

- the supernatural world;
- natural material objects, phenomena and processes;
- artificial material systems (technics, various constructions, roads, life subjects ...);
- artificial ideal systems (symbols, signs, knowledge, languages, etc.).

The optimal interaction of the man with these components is only possible when their features are known. In a general sense, "knowledge" is a certain notion of something in the man's consciousness (including consciousness itself). Scientific knowledge is a special kind of knowledge, which is characterized by the possibility of his comparison (as an ideal system) with some reality. Just for this reason, specific systems of knowledge may be true, objective or subjective, full or incomplete, exact or ambiguous, etc.

The appearance and development of science is connected with scientific knowledge. Versions of the definitions of the concept of science can be rather different:

- science is a form of reality display in public consciousness;
- science is a system of knowledge whose elements are philosophical and world-outlook bases and conclusions;

- science represents a special sphere of the purposeful human activity whose primary goal is studying objective relations in nature and society and establishment of laws which they obey in their development;
- science is a form of human knowledge which is a component of the spiritual culture of the society;
- science is a historically developing system of knowledge of the properties and relations of studied objects (nature, society, and human thinking).

More specific considering science allows resolving few directions:

- formation of a logically consistent knowledge system of the objective picture of the world;
- creation and elaboration of special means of cognitive activity (mathematical methods, research equipment, etc.);
- isolation of a special kind of professional activity (scientists, engineers, etc.) in the sphere of public labor division;
- functioning of a special system of organizations and establishments which are engaged in obtaining, storage, distribution, and introduction of the knowledge obtained (libraries, information centers, etc.).

First forms of prescience are associated with craft elements (from the Greek *teche*) arisen from empirical experience. To designate scientific activity, the term "wisdom" (from the Greek *bofos*) was originally used. The Egyptian priests, Iranian magicians, Babylon magicians, and first Ancient Greek thinkers were named "sages". Later, there appeared a notion of "lover of wisdom" (from the Greek *filisofos*) which designated the people engaged in speculative cognition of nature during the antique epoch. Aristotle named them "physicists" (from the Greek *fisis*, meaning nature).

The terms "science" and "scientist" were formed by the first half of the 19th century in the European university practice. They designated the activity in the sphere of mathematics, physics, chemistry, and other natural sciences. The term "social sciences" appeared a bit later.

Therefore, the following system of sciences had been formed by the 20th century:

- natural sciences as a system of the sciences of nature;
- social sciences as a system of the sciences of the man and society;
- engineering sciences as a system of the sciences most closely related with the realization/implementation of theoretical knowledge.

There are also other groups of sciences (e.g., medical, agricultural ones) which place emphasis on separate spheres of human activity. Interdisciplinary and general-scientific forms and cognition means are formed as well. At the same time, it is impossible to represent natural, public and engineering sciences as absolutely independent divisions of science. Analysis of the interaction among natural, public and engineering science shows that there exist interrelations among them and their borders are not so sharp. Besides, inside public, natural, technical and other sciences there are kinds and forms of activity which also interact among themselves, and the adequacy of the results of cognition process depends on the degree of this integration. However, integrative possibilities of the activity approach do not exclude isolation, in its frameworks, of various kinds, components, aspects, and characteristics of activity. E.g., science as a uniform system of activity appears consisting of research, scientific-organizational, scientific-information, scientific-pedagogical and scientific-auxiliary activity kinds among which there are also certain interrelations. Science development is objective process for which orientation to internal conditions, immanent (from the Latin *immanentis* – peculiar, inherent) for science. At the same time, the formation of natural sciences, humanities, and engineering in the increasing degree reveals its historical dependence on external conditions (social, economic, cultural ones, etc.)

Natural sciences are a subsystem of science. And in this sense, it is interrelated with all the elements of culture (religion, philosophy, ethics, etc.). On the other hand, natural sciences form an independent area of knowledge with its own structure, subject, and methods. Natural sciences are the area of human activity connected with production, ordering, and development of objective knowledge of natural phenomena and processes. It includes both cognitive activity as obtaining knowledge (observation, experiment, etc.) and its result as a system of scientific knowledge providing an insight into the natural-scientific picture of the world. The natural sciences purpose is description, analysis, and forecasting of natural phenomena and processes. Historically, all objects of inanimate and animate nature act as a subject of natural sciences. According to it, the natural sciences studying the objects of the inanimate nature (astronomy, mechanics, physics, etc.) and the natural sciences researching animate nature (biology, medicine, psychology, etc.) are traditionally resolved.

On the boundary of the 19–20 centuries, "butt" natural sciences (biophysics, chemical physics, biochemistry, etc.) appeared, formed in the course of interrelation of related subjects. Later, already in the 20th century, "synthetic" natural sciences (computer science, cybernetics, etc.) using methods of natural, technical and social sciences were formed.

Besides, it is quite natural that any science includes a set of methods and ways for studying the objective world.

The natural-scientific culture includes the knowledge of natural and artificial objects, phenomena, and processes; this knowledge is characterized by a certain degree of objectivity and reliability. The humanitarian culture is mainly defined by the social position of the man and includes the knowledge connected with social sciences and humanities. At the same time, both components of culture are interrelated, since both cultures represent the higher form of human knowledge, are elements of the uniform knowledge system, mutually coordinate the spiritual activity of the man, and promote the appearance and development of new interdisciplinary scientific leads.

It is possible to consider the following as the basic features of natural sciences:

- *universality*, caused by all mankind using the results of natural sciences;
- *fragmentariness*, expressed as studying various material objects and phenomena, and as natural sciences differentiation on different branches as well;
- *systematicity*, due to any natural science having a strictly certain structure;
- *continuity*, shown as the dependence of new knowledge on the previous one;
- *reproducibility and reliability*, directly determined by the possibility of experimental verification of any scientific conclusions;
- *rationality*, expressed in representation of obtained research results as concepts, theories, and laws.

Natural scientific knowledge is directly connected with such notions as scientific cognition, the object and subject of science, scientific language. Scientific cognition is usually represented as the process of knowledge acquisition. The object of science is what this science studies, while the subject is one who studies this object. Scientific language can be of either natural or artificial character. The natural scientific language reflects the properties, structure, and functions of a material object, and features of the mechanism of this or that phenomenon or process. The artificial scientific language is expressed in signs, symbols, mathematical, physical and chemical formulae and equations. Every natural science (or its section) has its own system of notions representing its language. E.g., they are mass, energy, velocity, wave, pulse, force, field … in physics; element, valency, chemical bond, acid, salt … in chemistry; organism, life, metabolism, live substance, gene … in biology; number, function, equation, differential, integr al … in mathematics; landscape, relief, climate … in geography, etc.

Unlike religion, reality and the laws established rather than belief are the basis of scientific knowledge; unlike art, the natural scientific knowledge does not stop at the level of images but draws the results of research to the development of a concept, the creation of a theory or formulation of a law; unlike ideology, natural sciences have general significance and do not depend on the social political system.

Besides, on the basis of the aforesaid it is possible to draw conclusions of principal differences of the true science from a pseudo one. The ethics of scientific research includes the freedom of scientific cognition, the establishment of scientific truth (this notion will be discussed below), the possibility of criticism of both initial and end results, the novelty of scientific knowledge. Pseudo sciences (whose typical representatives are astrology, parapsychology, ufology, and deviant science) are distinguished by their noncritical approach to the initial data and to the end results, infringement of the usual ethical standards, the actual absence of theories and laws explaining the properties of material objects and the mechanisms of phenomena or processes.

On the boundary of the 19–20 centuries, science became the determining factor of the development of socio-cultural processes of the modern civilization. These are science and the forms of its realization which determine the specificity of mutual relations of the man, society, and environment. By the beginning of the 20th century, the level of natural scientific knowledge allowed the man to become the force/power of "planetary scale" (Vladimir Vernadsky). However, this force/power is used by him not always reasonably. By the middle of the 20th century it became clear that, in essence, the realization of a rather great number of fundamental laws and principles of natural sciences leads, to some extent, to negative consequences for both the man and the environment of his dwelling. In a natural-scientific context, the Hamlet question is formulated as follows: science is good or evil? Its development is a blessing or threatens the existence of the man, society, and biosphere? Whether these questions mean a crisis of science?

In the history of science it is accepted to resolve two types of scientific knowledge. Namely: the Apollo-type science and the Faust-type one. In the former case, the science of antique type with its contemplation, passivity, locality, and irrationality is meant. In the latter case, it is the modern science with its activity, dynamism, globality, and rationality. Just with these characteristics, our representations of the "crisis" of scientific knowledge, the "deadlock" orientation of its development are associated. The activity of science turns as strengthening of its negative consequences, rationality demands irrationality as a complementarity, etc. Overcoming of the crisis tendencies in science development is connected with the necessity of inclusion of some new tendencies (humanization, ecologization / environmentalization, computerization, cosmization, etc.) into its structure. These tendencies are distinctly shown in the dynamics of scientific and technical progress (STP) and its modern form, namely, the scientific and technical revolution (STR). The STR is a stage of STP and socio-cultural development of the civilization, associated with qualitative changes in the "man–science–technics" system, with computerization of scientific and technico-technological processes, the use of biotechnological systems, perfection of administrative structures, etc. The dynamism of STR, the scale of the industrial-economic and socio-cultural activity of the society turn as strengthening of the negative factors associated with the realization of discoveries in fundamental sciences. This tendency is characteristic of the modern system of natural, social, and engineering sciences.

Within the limits of STR, its subsystem, "ecological revolution" (ER), focused on overcoming of the crisis situation in the "man–science– technics–society" system developed in the 1970-80s. On the boundary of the 1980–90s, STR is realized as "computer revolution" caused by large-scale microelectronics introduction into the industrial-economic and socio-cultural activity of the society. The traditional ways of obtaining, processing, and transfer of information change. Computer revolution (CR) has led to the process of informatization of the society. And ER and CR are in interrelation and interdependence. ER promotes reorientation of the traditional directions of civilization development; CR creates the necessary information and technico-technological conditions for rationalization and optimization of the man–society–biosphere relations. At the end of the 20th century, a tendency is recealed to overcoming of the "rupture" between natural scientific and humanitarian knowledge, the sciences of nature and of the man. On the one hand, the degree of science "culturizations" raises, i.e. its interrelation with the socio-cultural processes of reality amplifies. On the other hand, the process of "scientification" of culture amplifies, caused by penetration of scientific ideas, concepts, and representations into the set of knowledge of the man and society.

The modern scientific picture of the world increasingly gets system-integrative character, guided by the synergic approach based on cybernetic representations, the general theory of systems, self-organizing and self-management of systems of various levels. Just within its frameworks, preconditions for "carrying over" of the basic notions and representations from the sphere of natural sciences into that of humanitarian knowledge, and vice versa are generated. Natural-scientific and socio-cultural processes are considered in dynamics of their changes and interactions. Synergism is the basis of evolutionism in the widest sense of this word understood as the process of self-organizing of the Universe. Preconditions and conditions of the formation of the complete picture of the world is spoken of, which the modern scientific knowledge aspires to. Science thereby overcomes its crisis tendencies. The "crisis" of science in general and of natural sciences in particular stimulates their reorienttation to the directions providing sustainable development of both natural and socio-cultural systems.

1.2. General Forms and Methods of Natural-scientific Cognition

The sensual and rational forms of cognition are considered as most widespread. Sensation, perception, and representation belong to the former one while notion, judgment, and conclusion do to the latter one. Explanation, understanding, and prediction belong to the forms of cognition which have more general character and are original functional elements of scientific method. Explanation, as an element of the research process, is mainly used in natural sciences. The term "*understanding*" (which is also an original method of research) is more widespread in the social and humanitarian sciences, it is a way whose help the phenomena and processes of both individual spiritual life and public socially-humanitarian activity are usually interpreted with. As to understanding of nature, such a problem cannot exist in principle since the natural phenomena immediately have no definite purpose, i.e. they lack any sense introduced by someone in advance. *Prediction*, or *forecasting*, is close to explanation by its logic structure and is usually expressed as statements about possible facts, but, as a rule, on the basis of some theories or laws known before. Intuition plays a special role in this case.

Despite of essential distinctions among all sciences (including humanitarian ones), they are united by two circumstances. First, there are two basic forms of matter, namely, *substance* and *field*, which the microlevel, macrolevel, and megalevel correspond to. Features of matter at the *microscopic level* (atoms, atomic nuclei, elementary particles) are studied by modern physical methods. *The macroscopic level* is macrobodies, from colloidal particles (10^{-7} m) to such objects whose sizes the man constantly contacts with in the course of his life (millimeters, centimeters, meters, kilometers) up to the sizes of the Earth. Macrobodies consist of sets of Microsystems, which leads to essential qualitative and quantitative changes in their behavior. Substance of the macrolevel can exist in four states of aggregation of matter, namely, gaseous, liquid, solid, and plasma. The *megaworld* is full of space-scale objects (comets, meteorites, asteroids, planets, Solar system, stars, so-called black holes, galaxies and their congestions). It is quite natural that all the three structural levels of matter are interconnected and make a uniform continuum, but, nevertheless, each of them obeys its "own" laws; in particular, if studying them by means of physical sciences, these are the laws of quantum mechanics, classical physics, and general theory of relativity, respectively.

The second circumstance to unite all natural sciences consists in being one general sign признака to start with in any research, namely, the presumption of the existence of the object of cognition. This means the impossibility of obtaining reliable knowledge of that the possibility of whose existence has not been confirmed one way or another. Hence, the assumption of the existence of an object, phenomenon or process must be the first precondition of their research. In this connection, a new notion appears related with the process of cognition, namely *science methodology*. From the general viewpoint, science methodology formulates principles of the construction of forms and ways of scientific cognition, studies the properties of scientific knowledge, mutual relations among different systems of knowledge, and the regularities of their appearance and development. Therefore, at studying any object, the methodology of any science should use both specific research methods and general forms and ways of cognition. Generally, the notion "*method*" can be presented as a certain set of operations of the theoretical or practical cognition of reality.

Objectivity at establishing those or other features of the real world is a characteristic feature of the *scientific method.*

As most general methods of natural-scientific cognition the following ones are applied:

- *analysis* and *synthesis* as processes of mental or actual decomposition of an object or phenomenon as a whole onto its components (signs, properties, mechanism stages, etc.) and association of the studied parts into a whole system, respectively;
- *abstracting* as resolution of the principal properties and relations of an object under study and neglecting the insignificant ones;
- *generalization* as logic process of establishing new general properties and signs;
- *induction* and *deduction* as making conclusions from the particular to the general and from the general to the particular, respectively (on the basis of logic laws);
- *formalization* as a method leading to the creation of a certain language of science, i.e. to display of cognition results in most exact concepts and statements;
- *analogy* as a method in which on the basis of the similarity of some objects by some signs a conclusion is made about the similarity by other signs;
- *system analysis* includes examining the set of all interconnected, interdependent, and interacting objects, phenomena, and processes intended for the performance of a certain function;
- *functional method* is related with its main task, namely, establishment of basic functions of the object under study;
- *modeling/simulation* as the design of a conceptual model of the object under study.

From a general viewpoint, a *model* of the given object of research is another object which is compared сопоставляется with the initial one by its properties and features. It is also possible to consider as modeling the reproduction of the characteristics of some object on another one specially designed for their studying. Hence, anyway, a model reflects, in a preset way заданным образом, the chosen properties of the initial object of research.

Depending on a specific field of natural sciences, modeling has a certain sense, i.e. mathematical, physical, chemical, and other kinds of modeling exist.

Mathematical modeling is exploring the properties of some object by studying it mathematical model which is usually a set of mathematical equations to describe the behavior of this object. The mathematical model enables predicting this or that behavior of the object under changing conditions of its functioning. Besides, depending on the initial information of the object of modeling and the ultimate goal of research, different, by their form and structure, mathematical descriptions of the model are applied. Stochastic and statistical models are most widespread.

Stochastic models are designed on the basis of probability representations about the processes in the studied object and allow predicting its behavior by calculating probability distribution functions to characterize this or that property. Usually, stochastic models are used for analysis of functioning of objects under the conditions of casual events, for solving sophisticated problems (challenges) of long-term planning of a big enterprise, for analyzing possible consequences of unpredictable emergencies and process equipment failures, for establishment of most effective technological modes of some manufacture as a whole.

Statistical models are designed on the basis of some experimental data obtained at studying the object of research, and represent systems of equalities which relate the initial (input) and final (output) characteristics (parameters) of the object. The basic stages at statistical model design are the calculation of the parameters of the model, evaluation of the importance of the found values of these parameters, and verification of the conformity of the designed model to the studied object. The advantages of statistical models are the possibility of their application to objects with unknown mechanisms of the processes inside, and to big systems whose mathematical description would cause serious difficulties.

Physical and chemical modeling is usually reduced to studying models which differ from the initial objects by scale (e.g., laboratory and industrial plants). One of the most important conditions of physical and chemical modeling is the equality of the so-called *criteria of similarity* in the object and its model. These are certain dimensionless combinations of various physical and chemical quantities which directly influence the parameters of the object and its model. The efficiency of such kinds of modeling is determined, first, by the possibility of studying objects with lower expenses of substances, energy, and time; second, the possibility of exploring such objects whose physical and chemical properties and mechanisms of processes are not practically studied; third, carrying out much more complex measurements on the model than is possible directly on the object of modeling.

1.3. Significance of Theory and Experiment in the Structure of Natural Sciences

Theory is an important component of the process of cognition, whose overall objective is to explain empirically found regularities, their complete description, and prediction of new regularities. The theory's principles and laws are basic elements of any theory. The principles play the role of primary preconditions of the fundamental positions of the theory, while the laws, concretizing these principles, establish an interrelation of the consequences following from them. The possibility of getting knowledge of the object without direct contacting it is a feature of a theory. For this reason, any theory should be supplemented with a logico-gnoseological method opposite to abstraction, i.e. the corresponding interpretation. In spite of the fact that various theories are one of the steadiest forms of scientific knowledge, they enough often undergo both qualitative and quantitative changes owing to new empirical data accumulation. Such changes can a; so concern the fundamental principles of a theory; and transition to a new principle, as a rule, leads to a new theory which, in turn, should satisfy to Bohr's principle of conformity. According to this principle, every new, more correct and general theory should not reject the previous/preceding one but only be reduced to it under those conditions at which it has been constructed earlier (the development history of natural sciences knows many similar examples).

At the same time, it is necessary to note that, in general, theories are needed, first of all, for solving those problems which essentially cannot be solved without them. For example, the knowledge of the theory of algorithms is necessary to design new information technology and computers, the establishment of universal laws of nature is impossible without the knowledge of the theory of elementary particles, etc. Just for this reason, in every specific case there is a so-called subject domain of a theory which relates various hierarchical levels of the organization of the material world with the corresponding sciences, namely, the structural level with mathematics, the material level with physics, the techno-intellectual level with computer science, the technobiosystem level is related with the design of artificial biological systems (biotechnology and gene engineering methods). From here one of the general concepts of modern natural sciences follows, which states that the world is uniform while various sciences and associated theories study various/different aspects of inanimate and animate nature.

From our analysis of the general forms, methods, and significance of theory in natural-scientific cognition it is possible to draw certain conclusions on its structure. An empirical fact is the starting moment of any scientific research. Then, it is necessary to evaluate the frequency or probability of display of this fact. For this purpose, it is necessary to use special techniques of research whose important elements are observation and testing of the object under study, i.e. what is usually named "conducting an experiment". And such an experiment may be mental (based on some representations), modeling (based on studying some subjects or phenomena with their sizes, as a rule, proportionally reduced), and real (representing the facts). On the basis of the obtained experimental data, so-called empirical generalizations should be made which can allow the researcher to put forward some hypothesis. It should be noted that such a hypothesis often falls outside the limits the facts which have been its basis. Therefore, after advancing a hypothesis, it is necessary to come back to the experiment again (as a rule, at a higher qualitative level). And only in that case when the hypothesis passes repeated experimental tests, it can get the status of a regularity, concept or even law. Finally, all this can be a basis for the development of a theory to describe the specificity of the studied object. It is therefore possible to schematically present the structure of scientific cognition as follows: an empirical fact – observation – a mental experiment – a modeling experiment – a real experiment – empirical generalization – formulation of a scientific hypothesis – its verification by experience – introduction of terms and signs – obtaining quantitative relationships –creation of a theory.

So, the empirical and theoretical levels of cognition are in interrelation, causing each other. On the one hand, experiment is a necessary condition of the formation of an adequate theory. On the other hand, it is the theory which puts constructive problems before experimental science.

1.4. Basic Principles of Natural-Scientific Cognition

Basic principles of fundamental and applied sciences are connected with the structure of scientific cognition. Usually, the information model of obtaining scientific information is used to distinct fundamental and applied sciences. If ready knowledge forms and the object of cognition are taken as extreme "poles", a basic research is characterized by information movement from the object to the forms of knowledge which are the result of the research. Unlike this, an applied research is characterized by scientific information movement from the earlier obtained knowledge forms to the object of its application, use, introduction. Speaking of science as a whole and information movement in it and from it, it is very important to pay attention to the administrative function inherent in it in relation to manufacture and other practical spheres, which represents itself as one of the basic functions of information movement in the processes of science-manufacture integration. he administrative function is also somewhat inherent in fundamental sciences, but it is more distinctly expressed in applied sciences since it gets more specific character and realizes quite defined (including social) purposes here. Some fundamental sciences can use the apparatus of other fundamental sciences which acquire some traits of applied knowledge in that case. Despite of the basic distinction between fundamental and applied problems, there are general principles underlying any research process and based on the theory–practice interrelation.

One of such principles consists in that a theory can be only true after acknowledgement of its positions and conclusions by experience. Such an experiment must be performed in such conditions which would exclude the influence of extraneous (i.e. not concerning the object of research) factors and thereby does not distort the actual pattern of the phenomenon, process or the properties of the material object. The establishment of applicability borders of the given theory is an important stage in verifying the theory by experience. This stage also depends on objective and subjective factors. The objective factors are time (which is connected with the dynamics of the properties of the environment surrounding the man) and imperfection of the technique of experiment. The subjective factors are sense organ features and differing mental abilities of the man. Hence, this principle can be shortly formulated as follows: *experience is the criterion of truth.* Another basic principle of fundamental and applied sciences is *causality* (the "causality principle"). The following statement by Democritus is known: "Everything that happens has a natural cause, a cause that is inherent in the thing itself." From the positions of modern natural sciences, this principle means the presence of a cause-and-effect relationship among separate conditions of matter in the processes of its movement and development. *Causes* are the bases for the appearance and development of material objects, phenomena, and processes, including changes of their properties and mechanisms. These changes are called *consequences.*

The third principle is *the relativity of scientific knowledge*, i.e. scientific knowledge (notions, ideas, concepts, theories, models, etc.) is always relative and limited. This principle is associated with one of the primary goals of the researcher, namely, to find a so-called interval of discrepancy, i.e. the error of the theory or experiment which is always presented in any research.

The above principles are directly associated with the question of truth as of the main subject of cognition, since, in general, one of the basic questions of natural (and social) sciences is: in what measure one can trust scientific results and what should be considered as *scientific truth*? In this case, it is pertinent to remember Galileo's words: "the science Problem consists not in answering a question "as?", and in answering a question "how much?" (it was meant how much this or that result is close to truth имелось в виду – насколько тот или иной результат близок к истине). Hence, without going into further details of this problem, it is possible to draw the following conclusions concerning the definition and features of scientific truth:

- scientific truth is reproduction of material objects and phenomena so how they exist out of and irrespective of our consciousness;
- scientific truth is an adequate reflection of objects and phenomena by the cognizing subject;
- scientific truth is objective by content but subjective by form;

- any scientific truth is relative and contains an element of absolute knowledge only;
- the absolute truth (as an abstract notion) represents the sum of relative ones.

Besides, it is necessary to distinguish such notions as scientific truth and scientific outlook, which are not synonyms. *Scientific outlook* is such an objective relation to the material world at which every phenomenon finds its explanation not contradicting the main principles of natural-scientific cognition. The base of scientific outlook is the *scientific method* consisting in the following. First, the scientific approach to solving a posed problem is underlain by the use of the general methods of natural-scientific cognition (which are specified above). Second, the main task of the scientific method is strictly objective estimation of the studied objects and processes and generalization of the results obtained. Third, the initial precondition of any research is the statement that the material world represents an ordered and cognizable essence. Just in this connection, science depends on neither the epoch nor the public and political system, individual distinctions; and so it sharply differs from religious and other spiritual manifestations of mankind. Therefore, the true cognition of nature occurs when contradictions appear between theory and experience, which promote deeper cognition of nature and further development of the theory. Moreover, the stronger such contradictions are, the more radical is changing the laws describing the properties of natural and artificial material objects.

A special place in world cognition is occupied by *science methodology* as a doctrine of the methods of scientific cognition and transformation of reality, which includes the activity to produce and apply knowledge. Just for this reason, science methodology has a complex and integrative-interdisciplinary character.

1.5. Concept of "Two Cultures", The Philosophy and Natural Sciences Relationship

Despite of the raising status of integrative tendencies in natural sciences, the English writer and physicist Charles Percy Snow (1905–1980) formulated a thesis of the danger of contraposition of natural-scientific and humanitarian cultures in the middle of the 20th century. It was a question of that the development of natural sciences and humanities did not reach the desired level of integrativity. Scientists' opinions on that score disperse. Some ones fix a further "rupture" between the natural-scientific and humanitarian knowledge, predicting the approach of an epoch of "plurality" of cultures and strengthening of the differentiation tendency of knowledge of nature and the man. Some others note the raising status of the humanitarian-profile disciplines in relation to the sciences of the natural and technical cycle. And some others prove that the historically developed subordinated position of the humanitarian disciplines in relation to the sciences engaged in revealing natural laws remains. If more than thirty years back, when the polemic on the "two-culture" problem started, the question was, in essence, of the absolute domination of the status of natural sciences in the hierarchy of modern knowledge, by the end of the 20th century the scientists more often address such laws which traditionally come to light in the framework of the humanities. And this addressing not only is associated with searching for the answers to some questions of the philosophical, world-outlook or social character but also directly touches upon the sphere of their professional activity: revealing the "limits" of biological sciences penetration into the "secrets of life" or analysis of the socio-cultural consequences of the development of the modern orientation of physical knowledge. That the interrelation and interdependence of natural sciences, technical sciences, the humanities and arts amplifies is obvious. It does not mean there being no distinctions between them, which have real bases. If in the past the distinctions between them were absolutized, by the end of the 20th century, the elements uniting them are emphasized more and more.

The united, or "third culture" is reaching such a level of the socio-cultural development of the civilization at which the unity and interrelation of natural sciences, technical sciences, and the humanities comes to light. The reality of the "third culture", i.e. possible realization of integrative tendencies in science, is supported by the development level of modern knowledge. First, the traditional differentiation of scientific knowledge, characteristic of natural, technical, and humanitarian sciences, has prepared the basis for interdisciplinary interrelation of the developed system. Second, the apparatus of modern scientific knowledge is actually adapted for realization of integrative representations which are caused by the internal logic of science formation, the universality of structures and methods of scientific thinking. Third, the global problems common to all mankind and arising within the limits of civilizations at the end of the 20th century, require, for their resolution, activization of the processes associated just with integrative tendencies in the structure of science. The transition from the subject-disciplinary to problem-disciplinary development of science creates theoretico-methodological and social preconditions for constructive realization of integrative tendencies in the system of modern natural sciences, technical sciences, and the humanities. Reaching the level of the "third culture" assumes mutual enrichment of the principles and methods of the objective reality reflection, inherent in natural-scientific, humanitarian, and art knowledge. Conceptual thinking dominating in science and figurative образное thinking determining the style of art creativity, form an interconnected unity, expressing the indissolubility and interdependence of science and art. The "third culture" is underlain by the representation of the "unity of the world", which finally causes the unity of scientific and humanitarian knowledge. Therefore, in the framework of the "third culture", the unity of scientific knowledge is reached not by negation of the specificity of its various fields but is expressed in a variety of their forms.

Historically, philosophy (as a form of cognition of reality) and science are interrelated and are guided by solving a similar problem, namely, revealing regularities of nature and the essence of the man. Up to the middle of the 19th century, philosophy mainly acted in the form of "physiophilosophy", trying to interpret the "essence" of various things and reality phenomena in their "cogitative" integrity. The developing philosophical systems aspired not only to formulate the theoretical bases of world outlook but also pretended for the status of the "higher step", for the role of the "science of sciences". However, just the development of special sciences has had basic influence on philosophy which has kept its purposes and high status in culture.

The interrelation of philosophy, natural sciences, and other particular sciences is realized in several directions. First, philosophical analysis of various areas of science, interpretation of its achievements and revealing their possible philosophic-methodological consequences; second, the use of the philosophic-methodological apparatus (categories, laws, principles, etc.) for analysis of scientific knowledge, reaching a new level of theoretical knowledge; third, perception of the achievements and results of natural-scientific cognition by philosophical knowledge. During several historical periods, the interrelation between philosophy and science was expressed variously. During the antique period, the representation of the features of nature had the character of physiophilosophy, i.e. their mainly speculative interpretation. Philosophy was treated as the "science of sciences", offering its aprioristic schemes of cognition to other sciences. The process of science–philosophy splitting began in the Renaissance. On the one hand, the object of philosophy was as though narrowed, some sections becoming independent sciences (mechanics, physics, etc.) detached themselves from it. On the other hand, it was extended since sciences needed their philosophic-methodological substantiation and fundamental comprehension of the results. In the conditions of the Modern Era, the differentiation of natural sciences proceeded, the division process came to the end, but, at the same time, the interrelation of science and philosophy became stronger and passed to a more fundamental level. Now, philosophy not only gets from natural sciences empirical material for its own constructions but also promotes overcoming of the theoretical contradictions arising in the system of developing scientific knowledge.

The end of the 20th and the beginning of the 21st centuries show an increase of the status of philosophy in the hierarchy of modern scientific knowledge, the inconsistency of the representations about its "death". Just philosophy introduces the necessary humanistic element into the development dynamics of natural sciences, without allowing science to reduce to scientism, allows to comprehensively estimate the current state of science and to reveal basic directions of its development. At last, under the aegis of philosophy the process of overcoming of the rupture between the "two cultures" proceeds.

REFERENCES

Ananiev. B.G. *Systematic Approach in Modern Science*. Moscow, 2002. [Russ].

Bernal J.. *Science in History* (The Emergence of Science; The Scientific and Industrial Revolutions; The Natural Sciences in Our Time; The Social Sciences: Conclusions). Faber and Faber, 1954.

Einstein A.. On Science and Religion. Edinburgh, Scottish Academic, 1940.

Fenberg E.L.. *Two Cultures*. Intuition and Logics in Art and Science. Moscow, 1992. [Russ].

Gumilev L.N.. *Ethnosphere*. The History of People and the History of Nature. Moscow, 1993. [Russ].

Kuhn T.S.. *The Structure of Scientific Revolutions*. University of Chicago Press, 1962.

Kuznetsov V.I., G.M. Idlis, V.I. Gutina. *Natural Sciences*. Moscow, 1996. [Russ].

Philosophical Problems of Natural Sciences. Moscow, 1985. [Russ].

Poincare H.. *The Foundations of Science*. New York, Science Press, 1908.

Shklovsky I.S.. Universe, Reason, Life. Moscow, 1987. [Russ].

Stepin V.S.. *Philosophy of Science*. Moscow, 2006. [Russ].

Vashchekin N.P., V.A. Los, A.D. Ursul. *Concepts of Modern Natural Sciences*. Moscow, 2000. [Russ].

Chapter 2

MATHEMATICS AND NATURAL SCIENCES

One, who wants to solve questions of natural sciences with no mathematics involved, poses an unsolvable problem. Measure what is measurable, and make measurable what is not so.

Galileo Galilei

2.1. SPECIFICITY OF MATHEMATICS AND BASIC STAGES OF ITS DEVELOPMENT

From the very instant of its origin, mathematical knowledge showed a tendency to generality. Pythagoras and his followers either identified numbers with things themselves or saw the essence and causes of things in numbers. A divine-religious sense was given to numbers; they were elevated to the rank of the main principle of the world's existence. The idea of the universality of mathematics and future scientific knowledge mathematization was contained in the concept of the Pythagorean Union. The idea of the generality of mathematics occupied Descartes' thoughts as well, who believed that "there must be some general science to explain that element as a whole which gives rise to problems about order and measurement, restricted as these are to no special subject matter. This, I perceived was called 'universal mathematics' ...". This great philosopher and mathematician focused his attention on the possible general-scientific character of mathematical knowledge, ennobling it over other sciences. The idea of the universality of mathematics was further developed by Leibnitz, and the philosopher Kant considered that "a doctrine of nature will contain only so much science proper as there is applied mathematics in it". However, only in the second half of the 20th century, the successes of applied mathematics and modern computer science allowed reviving (on a new basis) and justifying the ideas of those thinkers who had spoken about the universal character of mathematics. Nowadays, when the process of mathematization has captured natural, technical, and social sciences (including fundamental and applied ones) and exhibited its productivity, the general-scientific character of mathematical knowledge has manifested itself definitely enough.

Let us consider basic features of mathematics as a general scientific phenomenon, and its integration and interactions with particular sciences on the way of knowledge mathematization.

What does mathematics study? Only recently this science was considered to study the quantitative relations and spatial forms of reality. This concept already isolates mathematics from other sciences studying a certain fragment of the nature, society, or thinking. Mathematics, in this view, studies a certain aspect of reality, not a fragment of it, which is immanently inherent in all forms of the movement of matter. Therefore, such an interpretation of mathematics points to this science as to a wide enough branch of knowledge, essentially distinct from other particular sciences.

Some changes occurred in mathematics in the last century, which have expanded the subject and methods of its research and isolated it from particular sciences. Mathematics differs, first of all, by its not focusing attention on the spatial and quantitative relations of reality only. It also investigates other aspects of both the external world and our thinking (the ideal world) and creates it as well as reflects it. We have every reason to consider that mathematics is defined not by its subject of research (which is continuously enriched and complicated) but by the method of its approach, representing merely a specific language for logical symbolic data processing in other sciences. However, it would be strange to count that mathematics has no subject and acts just as a method for other sciences. Mathematics really looks as a method (research means) for external use but inside itself it, nevertheless, is engaged in searching its own subject with the purpose to transform it into a method. What proves it is the mathematical knowledge evolution. The difference of mathematics from natural sciences is also due to that its axioms must be logically consistent, whilst the requirements of validity, obviousness, visualization, and experimental confirmation are inapplicable to it.

At consideration of the specific features of modern mathematics and at revealing its subject, method, and prospects, it is necessary to note three directions of development, which substantially define it as a science. It is, first of all, the modern dominating structural approach, i.e. the representation of mathematics as of a set of certain mathematical structures, based on the set-theoretic approach. Second, it is the category-theoretic approach developing in last decades, which has declared its claim for the status of the "principles" of modern mathematics. And though it is not recognized so widely but possesses a great ideological charge for mathematical knowledge progress. Third, it is cybernetization and informatization of modern mathematics, the development of discrete mathematics, and the wide usage (especially for applied purposes) of computers. The changes in modern mathematics have essentially affected our understanding of its subject. This, first of all, concerns the notion of quantitative relations, which the notion of mathematical structure is now included into, in addition to quantities and spatial relations.

According to Bourbaki's concept, the definition of the notion of mathematical structure is connected with specifying several relations which its elements are in, and with introducing conditions, i.e. such axioms which the introduced relations of the structure should satisfy. Mathematical structures express abstract general steady relations, segregating the nature of the initial elements and their interpretations. Nicolas Bourbaki has resolved three basic structures. *Algebraic structures* are most often sets, in which such a relation between three elements is specified when the third element is unequivocally defined by the two others. Groups, rings, and fields exemplify this type of structures. In *topological structures*, attention is focused on the notions of limit and continuity, and in *order structures* it is paid to orderliness relations. The structural approach in mathematics is further development of the set-theoretic approach to its subject and allows uniting all the existing sections of modern mathematics from general positions.

At the same time, structural set-theoretic representations cannot now pretend to the role of unique and most general bases of mathematical knowledge since they are not always necessary and effective. E.g., in relation to the definition of the concept of vector (usually defined as a directed straight-line segment), this approach is rather complicated. The category-theoretic concept essentially simplifies bulky set-theoretic constructions in mathematics. The term of category in mathematics is adopted from philosophy (Aristotle and Kant), which is due to the aspiration to introduce most general notions in mathematics as well. The mathematical theory of categories can be considered, in a certain relation, as further generalization of the theory of sets. The mathematical theory of categories is, to a certain extent, generalization of the set-theoretic approach but does not replace it completely. The theory of categories has arisen on the joint of two most general mathematical theories, as synthesis of the algebraic and topological structures. The further development of mathematics has shown that this new theory reflects something essential to modern mathematics, prepared by its previous development. This means that no appearance of the theory of categories would be possible without the «set-theoretic stage» of mathematics developments, without elaboration of algebraic theories, topology, functional analysis, mathematical logic, etc. In the course of development of these theories and, in general, the set-theoretic thinking in mathematics it has become clear that most important results in it have been obtained while researching the relations between objects rather than objects themselves.

2.2. Mathematical Method and Experiment

The notions of spatial forms, quantitative relations, mathematical structures, categories, etc. reflect essential aspects of mathematics as a science having not only its subject but its method as well. Mathematics as a method acts as some kind of secondary science directed on serving other sciences (e.g., physics) having a specific material object and stimulating mathematical knowledge development. This has created an illusion of mathematics being among sciences, where observation and experiment play a decisive role. Mathematics is not based on experimental facts and public practice as experimental natural sciences are; it can create its own virtual theoretical reality without them, it can precede any experimental theory or cannot find such one for a long time. However, there is one new form of practice where mathematics creates its own experiment. It is so-called mathematical experiment which has arisen on the way of scientific knowledge informatization and represents one of the forms of mathematics– cybernetics integration.

Mathematical experiment, as Victor Glushkov has noted, "takes an intermediate place between the classical deductive and classical experimental methods of research. Its appearance has led to a new philosophical problem, namely, the necessity of reconsideration of not only the subject of mathematics but also of the method of mathematical research". Long ago, mathematicians included a number of theories and disciplines dealing not only with spatial (geometrical) or quantitative characteristics into the structure of their science. This is due to the apparatus of deductive constructions being well developed for the objects studied by mathematical logics. In the case of the method of mathematical experiment, no similar apparatus is almost used. At the same time, the experience of science development shows that the process of knowledge mathematization inevitably leads, sooner or later, to the development of such an apparatus. Therefore, mathematics returns adequate possibilities of expansion of the deductive construction apparatus used by it for the huge qualitative variety of research objects in various fields of knowledge. The method of mathematical experiment, essentially strengthening mathematics expansion into other sciences, gives evidence of the appearance of new essential links between mathematics, cybernetics, and computer science, expanding and integrating their traditional subjects of research. Owing to the application of powerful computers, more high-grade mathematical modeling becomes possible for both internal and external needs since it is no longer necessary to simplify mathematical models for their usage in natural-scientific theories and at empirical data processing. By means of a mathematical experiment, one can spend considerably more computing procedures and design an optimum mathematical model to solve this or that scientific problem. This especially concerns global problems of mankind. Mathematical experiment in natural-scientific research has given a powerful imact to the development of all sciences about inanimate and animate nature, raised them on a qualitatively new level, promoted the elaboration of deeper theoretical concepts and obtaining new scientific knowledge. At the same time, the informatization of mathematical means has resulted in inclusion of new types of models, sets of programs and algorithms into the mathematical arsenal as well as in traditional method perfection.

2.3. Modern Problems, Leads of Mathematics, and Science Mathematization Limits

The development of mathematics and the mathematization of other sciences are interconnected processes. Mathematics as a system of mathematical structures and categories cannot develop without supplying substantial scientific knowledge with its means and logico-symbolic potential. In turn, the “substantial” sciences (first of all, physics, chemistry, and biology) positively influence mathematics as a science with its own subject and method. The application process of mathematical means in any field of natural-scientific knowledge makes the essence of science mathematization, which is of integrative character and generates new knowledge. Science mathematization thereby confirms the thesis of new knowledge most often appearing on the ways of scientific discipline interactions.

Mathematization as scientific knowledge synthesis, having its own content, manifests itself in various forms. First, it is expressed in transferring methods (used at systematization and formalization) into other fields of science. Second, it is shown as quantitative analysis and expression of qualitatively established empirical data with the purpose of their use for the formulation of numerical laws. Third, mathematical models or special mathematical branches of various sciences, e.g., mathematical psychology or econometrics, can be its forms.

At the present stage, knowledge mathematization substantially depends on the use of means, and it apparently represents the basic lead of mathematization. It obviously includes computer science with its methods and cognition means. The use of computers for faster and exact computing works, for measurement result processing, representing one of the major directions of mathematization, is its simplest display by means of computers. The genuine knowledge mathematization on the computerization way occurs only when computers are applied together with the corresponding difficult deductive apparatus and logic calculations, and also for searching for new laws, the design of a special formalized language of science, and construction of more fundamental theories.

Some restrictions of using formal mathematical methods follow from Kurt Gödel's theorems. Согласно первой теореме Геделя (теореме о неполноте) в достаточно сложных формальных системах всегда имеются недоказуемые и одновременно неопровержимые в данной системе выводы. According to the first Gödel theorem (the theorem of incompleteness), there always exist indemonstrable and, simultaneously, incontestable (in the given system) conclusions in difficult enough formal systems. Вторая теорема Геделя утверждает невозможность доказательства непротиворечивости формальной системы средствами самой этой системы. The second Gödel theorem states the impossibility to prove the consistency of a formal system with the aid of the means of this system. The first incompleteness theorem states that no consistent system of axioms whose theorems can be listed by an "effective procedure" (e.g., a computer program, but it could be any sort of algorithm) is capable of proving all truths about the relations of the natural numbers (arithmetic). For any such system, there will always be statements about the natural numbers that are true, but that are unprovable within the system. The second incompleteness theorem, a corollary of the first, shows that such a system cannot demonstrate its own consistency. These theorems point to the impossibility of full formalization of rather substantial processes and results of human thinking; they are usually treated as serious restrictions of scientific knowledge mathematization. At outlining the borders of mathematization, it is also necessary to mean that that the mathematical language has its natural limits of the knowledge content expression. In general, abstract thinking is itself limited if isolated from the external world; if this isolation is broken and there appears a link with the external world, these restrictions are lifted. Though mathematization is limited historically, its limits are mobile and conditional, and are finally caused by the unity of strict and nonstrict, formal and substantial methods. In the course of mathematization, the interaction of mathematics with other sciences and that of science as a whole with the external world in the course of practical activity will move apart the mathematical means usage borders more and more. Therefore, knowledge synthesis performed in the course of mathematization generates new possibilities of the application of mathematical means as well as of the enrichment of these means with new features.

The restrictions of mathematics usage in social sciences are due to mathematics, for the present, lacking rather developed means for effective applications in social knowledge. In the development of mathematics, its theories and methods, a special place is hitherto occupied by its closest interaction with the sciences of the inanimate nature, despite its principal general-scientific character. Nevertheless, the interaction between mathematics and the sciences about the inanimate nature and society strengthens at the present stage. Other restrictions are caused by the nature of the social form of movement and its extreme complexity in comparison with natural phenomena. There are much more objective obstacles for the mathematization of social studies in comparison with natural and technical ones. However, the greatest possibilities of mathematization of the humanities can be connected with the application of novel information technologies in these fields of scientific knowledge. It is also necessary to mean that, as well as any development of knowledge, mathematization is accompanied by certain negative consequences. The language and methods of mathematics are known to be sometimes used for proving false ideas and theories. Mathematical notions, owing to their abstraction from the specific content of the objects under study, may undergo certain manipulations with some purposes different from searching for truth. All these discussions of the mathematization limits point to the necessity of examining its efficiency and expediency, which assumes refusal of our simplified representation of it as of a good only, demands control of the process of mathematization to avoid its negative consequences and effects.

As to the positive direction of scientific knowledge mathematization represented itself as an integration-general scientific process, it, during its development, generates knowledge synthesis between various sciences, theories, and other forms and means of cognition. The integrative potentialities of mathematics are closely related to the nature of mathematical abstractions, which distract from the specific nature of any object under consideration. Therefore, when a certain mathematical theory is applied to a specific object of research (e.g., to message transfer process in technical communication channels), it cannot be applied to many other objects (to information transfer in genetic biological systems, between animals communicating among themselves, people, the man and car, etc.).

The possibilities of mathematical models for investigated objects are not usually restricted by the limits of the initial scope, so their scope of application gradually extends. Translation of substantial ideas previously related with the mathematical apparatus is also done through this apparatus which acts as though as a bridge with which help the movement of knowledge from one field of science to another goes with new ideas to be generated.

REFERENCES

Arfken. G. *Mathematical Methods for Physicists*. Academic Press, New York, London, 1968.

Arnold V.I.. *Mathematical Comprehension of Nature*. Moscow, 2009. [Russ].

Bohr N.. *Collected Works*. 13-Volume Limited Edition Set / Ed. Finn Aaserud, North Holland, 2008.

Bourbaki N.. *Elements of the History of Mathematics*. Springer-Verlag, 1994.

Glushkov V.M.. *Introduction into Cybernetics*. Kiev, 1964. [Russ].

Gurevich I.M.. *Laws of Informatics as the Bases of Design and Cognition of Complex Systems*. Moscow, 2007. [Russ].

Mathews J., R.L. Walker. *Mathematical Methods of Physics* / 2nd ed., Addison-Wesley Publishing Company, 1970.

Schwatzburd S.I.. *Mathematics and Natural Sciences*. Moscow, 1969. [Russ].

Ursul. A.D. *Informatization of the Society* (an Introduction into Social Informatics). Moscow, 1990. [Russ].

Yaglom I.M.. *Mathematics and Real World*. Moscow, 2006. [Russ].

Zorich V.A.. *Mathematical Analysis of Natural Scientific Problems*. Moscow, 2008. [Russ].

Chapter 3

BASIC STAGES OF THE WORLD'S PHYSICAL PICTURE DEVELOPMENT

We feel clearly that we are only now beginning to acquire reliable material for welding together the sum total of all that is known into a whole.

Erwin Schrödinger

3.1. A BRIEF CHARACTERISTICS OF PHYSIOPHILOSOPHICAL REPRESENTATIONS

Throughout many years (approx. before the mid-16th century), nature's features were explained on the basis of some general philosophical representations and principles united by the notion of physiophilosophy (nature philosophy). Its founders constructed their representations of natural objects, phenomena, and processes on the basis of hypotheses and assumptions. We will note several achievements of some representatives of physiophilosophy, most important for that time as well as which are used in modern natural sciences.

The majority of the historians of natural science development consider that approx. in the mid-6th century BC there appeared a first scientific-philosophical school founded by Pythagoras (584–500 BC). Perception of the world in the form of number was the basic concept of this school (later named as "the Pythagorean union"). Principles of arithmetics, the basics of the theory of numbers (in particular, the doctrines of even and odd, simple and compound numbers, divisibility of numbers, the notions of arithmetic and geometric means were introduced) and geometrical algebra were laid in the works of Pythagoras and his pupils. This was Pythagoras who coined the word "cosmos", presented the Earth as a sphere, explained the solar eclipse by the Moon passage between the Sun and the Earth, and explained the alternation of seasons by the terrestrial orbit inclination. Besides, the world culture is obliged to Pythagoras as to one of the founders of the theory of music, who mathematically described musical intervals (in particular, a third, a quart, a fifth, etc.) and presented the harmony of nature by the balance of such contrasts as limit and infinity, evenness and oddness, unity and many, right-hand and left-hand, rest and motion, straight and curve, light and darkness, female and male, good and evil. Therefore, the significance of the Pythagorean union in the history of natural sciences consists in the establishment of the unity of qualitative and quantitative variety of natural objects, i.e. the properties of every object and phenomenon have a certain measure expressed by number and certain harmony.

Democritus (460–370 BC) was the first to put forward the hypothesis of atomism, he is considered as the founder of antique materialism. In his opinion, the real world consists of existence and nonexistence. Nonexistence is emptiness which is motionless, infinite, and the properties of the bodies being in it do not depend on it. On the contrary, existence consists of independent smallest particles of substance, *atoms*, which neither arise and nor disappear; they are impenetrable, indivisible, neutral, and different in size and shape. Democritus' atomistic concept was one of most fruitful (from the viewpoint of matter structure representations) and led to the formulation of the determinism principle, which, on the one hand, identified causality and necessity, and, on the other hand, rejected accident existence.

Plato (428–348 BC) created "the mathematical program". He explained contradictions in nature by the presence of two worlds or two realities. The first (material) world is a set of constantly changing (in the human consciousness) substances. The second world is a set of general and invariable essences (ideas, notions, etc.), comprehended by mind rather by feelings. Plato, developing Pythagoras' representations, assigned the leading role in his concept to mathematics, considering number as the basic key to nature cognition. At the same time, in spite of the fact that Plato's physiophilosophy was penetrated by mathematical images and analogies, he, unlike Pythagoras, considered the environment (including cosmos) as non-uniform and divided it into the divine (the sky, etc.) and terrestrial ones.

Aristotle (384–322 BC) authored hundreds of works in the fields of philosophy, logic, psychology, politics, ethics, esthetics, and natural sciences. He was Plato's pupil, but did not agree with him in many respects. His work "De Anima" ("On the Soul") is hitherto considered as a classical work on psychology, and his book "Physics" allows us to consider him as one of the founders of physical sciences. However, the main place in his works is occupied with the doctrines of matter and form, and about motion as qualitative and quantitative changes of bodies' locations in space. According to the first doctrine, every separate thing is a combination of matter and form, while matter itself is chaotic substance. Unlike Democritus, Aristotle constructed the so-called continual world picture (opposite to the atomistic and discrete one). He divided mechanical motion into three types, namely, circular (whose cause is God and, consequently, it is inherent in all heavenly bodies), natural (motion of heavy bodies downwards and easy ones upwards), and forcible (motion of bodies under the influence of external forces).

An essentially new stage in mathematics development is related with the activity of Euclid (the first half of the 3rd century BC) who considered geometrical problems on a plane in his two fundamental works "Elements" and "Optics", created the bases of planimetry and stereometry, formulated some principles which subsequently became the basis in the development of geometrical optics.

Archimedes (287–212 BC) was an outstanding mathematician, mechanic, and engineer. The principle of the lever is one of his most important discoveries. He introduced the notion of "the moment of forces", constructed a model of the heavenly sphere, developed the bases of mathematical physics, calculated the π number (equal to 3.14), derived some expressions to allow calculating the areas of surfaces and the volumes of bodies of various geometrical shapes (a sphere, ellipse, cone, etc.), the sum of a geometrical progression, the location of the centers of gravity of various geometrical figures (a triangle, parallelogram, trapeze, etc.). Archimedes formulated the law underlying modern hydrostatics and aerostatics (published in his book "On Floating Bodies"): "Any floating object displaces its own weight of fluid" or "A body immersed in a fluid experiences a buoyant force equal to the weight of the fluid it displaces".

Analysis of the listed achievements of the initial stage of natural science development allows drawing a general conclusion which consists in the following. The ancient Greek physiophilosophy was based on the concept of the unity of all real beings, i.e., first, on the idea of all things originating from some uniform source (e.g., air, water, fire, etc.); and, second, on the idea of the general animateness of matter. The development of phisio-philosophers' representations occurred during the ancient and medieval periods (up to the late 15th–early 17th century) and, as appeared later, the general characteristic development of natural sciences during this rather long period was transition from the anthropocentrism concept to geocentrism and, then, to the heliocentric system of the world. The development of the main principles of physiophilosophy (and, in certain cases, refusal of them) is associated, first of all, with introduction of experimental research methods into natural sciences.

3.2. Mechanistic Concept

Another major stage of the formation of the natural-scientific picture of the world, much closer to modern natural sciences than physiophilosophy, is the so-called mechanistic picture of the world originating from ca. the end of the 16th century. The outstanding Renaissance artist Leonardo da Vinci (1452–1512) is considered as one of its first developers, who has been no less a great scientist. In the field of botany, he discovered the law of leaves arrangement on the stalk, developed a way to determine the age of a tree from its annual rings, and created an atlas of plants (used, in a certain degree, till now). In the field of anatomy, he described the right cordial ventricle valve, offered models of the brain, heart, and eye. Studying the regularities of geometrical optics, he presented an eye as a lens due to which a turned image of an object resulted. In geological research, sea adjournments and the formation processes of sedimentary rocks were described. In mechanics, Leonardo da Vinci pioneered by constructing models of a glider, caterpillar, and helicopter.

The Polish astronomer Nicolas Copernicus (1473–1543), who created the heliocentric system of the world, and the Italian scientist Giordano Bruno (1548–1600) contributed much into the development of representations about the mechanistic picture of the world. A mathematical theory of motion of the Sun, Moon, some planets, and stars is expounded in Copernicus' book "De revolutionibus orbium coelestium" ("On the Revolutions of the Celestial Spheres"). The possibility to transfer from geocentrism to heliocentricism was based on the idea of the relative character of motion. These were the representations about the mobility of the Earth and about the motion of all planets as a uniform system which became the major principles in Copernicus' doctrine. The world-outlook and methodological significance of the heliocentric concept created by Copernicus consists in its scientifically proving the features of many astronomical phenomena (including season changes, loop-like motion of planets, etc.), completely rejecting the representation of the geocentric system of the world, being one of the bases of classical mechanics.

One of Bruno's main proceedings is his monograph "De l'Infinito Universo et Mondi" ("On the Infinite Universe and Worlds") (1584). It states the infinity of the Universe, the uniformity of its structure, assumes the possible presence of numerous heavenly bodies (including the Sun and planets, self-shining and dark stars) in it, the existence of continuous energy and substance exchange between these heavenly bodies, the invariance of the primary material substance underlying all bodies and phenomena. Despite the incontestable achievements of these scientists, the creation of the mechanistic concept is mainly associated with such names as Galileo Galilei, Johann Kepler, and Isaac Newton.

The basic merit of Galileo Galilei (1564–1642) consisted in his first introducing such methods as observation, performing experiments, and quantitative data processing into the practice of nature studies. E.g., in 1609 he designed a 30X telescope with whose help he discovered craters and mountains on the Moon, observed stains on the Sun, discovered the Milky Way and four satellites of Jupiter. Besides, Galilei deduced a number of fundamental laws of mechanics (e.g., the speed equality of freely falling bodies). Unlike phisiophilosophers, Galilei introduced the notion of inertial movement, which meant that any physical system is either at rest or moves uniformly and rectilinearly (naturally, in absence of external forces' influence upon the body). At the same time, Galilei separated the notions of uniform and accelerated motion, defined acceleration as the speed of speed changing, showed that acceleration results from force action upon the moving body. At last, he first formulated the principle of motion relativity, whose essence consists in all inertial systems being equivalent among themselves in the description of mechanical processes.

The German mathematician and astronomer Johann Kepler (1571–1630) discovered the basic motion laws of the Solar system planets. The first law states that the planets move on elliptic orbits. The second law says that the motion of the planets on their orbits occurs non-uniformly, and the speed changes in such a manner that the areas described by the radius vector for equal time intervals are equal as well. The practical importance of this law consists in its allowing deriving an equation to calculate the location of any planet on its orbit at any preset instant. The third law establishes the dependence between the cycle times of the planets and their average distances from the Sun. Mathematically, this dependence is expressed so that the squared periods of rotation of the planets round the Sun relate as the cubed average distances of these planets from the Sun. The concept follows that the Sun is the basic power source driving the planets. From the idea of magnetism universality (a property inherent in the majority of macrobodies), Kepler hypothesized a mechanism of action of the force driving the planets; this mechanism is caused by the presence of a whirlwind with a very high speed of motion of the Sun's magnetic field arising from its rotation. Besides, Kepler explained ebb and flow of the oceans by the Moon influence, designed a theory and developed ways to predict solar and lunar eclipses, invented an optical system used in modern refractometers.

Isaac Newton (1643–1727) was a fine mathematician as well as an outstanding physicist. To solve the primary tasks of mechanics, Newton designed special variants of differential and integral calculus: i) speed was presented as the first derivative of the way (distance) with respect to time, and ii) acceleration is the first derivative of speed or the second derivative of the way with respect to time. Newton formulated the basic laws of dynamics and the law of universal gravitation in his monograph "Mathematical Principles of Natural Philosophy" (1687) which actually completed the construction of the mechanistic concept. *1st law*: an object at rest tends to stay at rest and that an object in uniform motion tends to stay in uniform motion unless acted upon by a net external force. *2nd law*: an applied force on an object equals the rate of change of its momentum with time. *3rd law*: for every action there is an equal and opposite reaction. *4th law* (the universal gravitation law): every point mass in the universe attracts every other point mass with a force that is directly proportional to the product of their masses and inversely proportional to the squared distance between them.

Proceeding from the physical sense of Newton's listed laws, it is possible to draw conclusions to highlight some prominent features of the mechanistic concept:

1) To describe mechanical motion, it is necessary and sufficient to know the coordinate of a body, its speed, and the equation of its motion at some instant of time. Under these conditions it is always possible to precisely determine the location of the body at any other instant of time (i.e. in both the past and future).
2) The majority of the regularities observed in nature can be explained by mechanical motion, and it is possible to present the whole environment surrounding the man as a global system whose future state can be unequivocally defined by its previous state.
3) Movement is simple mechanical moving, and the movement laws are fundamental laws of the universe.
4) Bodies move uniformly and rectilinearly, and the cause of deviation from such motion is the action of an external force upon them, which is connected with one of the properties of motion of the body, named inertia.
5) The interaction between bodies occurs instantly at any distance, i.e. the influence can be transferred in empty space and with an infinitely high speed; it manifests itself in the *long-range action principle (action at a distance)*, which was first introduced by Rene Descartes (1596–1650). It means that if one body acts upon another one, that second body is affected at the same instant. The gravitational interaction between macrobodies is rather weak (by energy) and it is often difficult to measure it. Nevertheless, Henry Cavendish succeeded to experimentally determine the gravitational constant ($6.6 \cdot 10^{-11}$ $N{\cdot}m^2/kg^2$) in 1797.
6) All mechanical processes obey the determinism principle, i.e. accident is excluded from the world picture.
7) The mechanistic concept corresponds to the discrete (or corpuscular) reality model. This means that matter is a material substance consisting of atoms or corpuscles which are absolutely strong, indivisible, impenetrable, and characterized by the presence of mass.
8) Space is three-dimensional, constant, and does not depend on any properties of matter; time also depends on neither space nor matter; space and time are not directly related with the motion of bodies, i.e. they have absolute character.

3.3. Electromagnetic Concept

Electric and magnetic phenomena were known for a long time, just in ancient centuries. However, the interrelation between these physical phenomena was established only in the 19th century as a result of the works of such scientists as the Dane Hans Christian Ørsted (1777–1851) and the Englishmen Michael Faraday (1791–1867) and James Clerk Maxwell (1831–1879). Such facts as a magnetic needle's deviation over a conductor with an electric current, electric current occurrence at rotation of a closed contour in a magnetic field, etc. pointed to such a relationship.

Table 1. Electromagnetic Spectrum

EM wave range	Frequency range, Hz	Wavelength range, m	Basic applications
Radio waves	$3\times10^5-3\times10^7$	10^3-10	Radio communication
Metric waves	$3\times10^7-3\times10^8$	$1-10^{-2}$	Television
Super-high frequency (SHF) microwaves (decimetric and centimetric waves)	$3\times10^8-3\times10^{10}$	$1-10^{-2}$	Radar location, outer-space communication, technological processes
Extremely high frequency (EHF) microwaves (millimetric waves)	$3\times10^{10}-3\times10^{11}$	$10^{-2}-10^{-3}$	Radar location, space, control systems, biology, and medicine
Terahertz (submillimetric) radiation	$3\times10^{11}-10^{14}$	$10^{-3}-5\times10^{-6}$	Communication, biology, and medicine
Infrared radiation	$10^{14}-34\times10^{14}$	$5\times10^{-6}-10^{-6}$	Location, medicine
Visible radiation	$3\times10^{14}-10^{15}$	$10^{-6}-5\times10^{-7}$	Everywhere
Ultraviolet radiation	$10^{15}-3\times10^{15}$	$5\times10^{-7}-10^{-7}$	Medicine
X-ray radiation	$3\times10^{15}-10^{18}$	$10^{-7}-3\times10^{-10}$	Technology, medicine
Gamma radiation	$3\times10^{-10}-3\times10^{-13}$	$10^{18}-10^{21}$	Technology, nuclear physics, and medicine

These and many other experimental data allowed drawing fundamental conclusions about there being, at least, one more (except gravitation forces) physical force field, namely, the electromagnetic field. The existence of other force fields different from the known ones by nature is possible, in principle.

Both electric charges and a variable magnetic field are electric field sources. Coulomb's law serves the basis of interaction between point electric charges: The magnitude of the electrostatics interaction force between two point charges is directly proportional to the scalar multiplication of the magnitudes of charges and inversely proportional to the squard distances between them.

Michael Faraday, James Clerk Maxwell, and Hendrik Lorentz (1853–1928) are the founders of the doctrine of electromagnetic field, they introduced the notions of the electric and magnetic field strength, described the phenomena of electromagnetic induction and self-induction, discovered the laws of electrolysis, pointed to the oscillatory character of propagation of electric and magnetic energy. It follows from the theory of electromagnetic field that its energy propagates in the environment as electromagnetic waves; and the speed of such propagation is the highest possible speed of energy transfer in nature and is equal to 300,000 km/s.

Therefore, the features of the electromagnetic concept can be characterized as follows.

1) Electromagnetic representations have allowed offering a field (or continuous) reality model. This means that:
   - matter is a uniform continuous field with point power centers which are electric charges and wave motions in this field;

- the surrounding environment is an electrodynamic system made of electrically charged particles interacting among themselves by means of an electromagnetic field.

2) Energy transfer by a field proceeds from point to point continuously and with a final speed, which manifests itself in *the short-range interaction principle*: the interaction of electric charges occurs not instantly, i.e. every electrically charged particle induces an electric field acting upon other particles not at the very instant but after a while. Therefore, interaction is transferred through an "intermediary", namely, the electromagnetic field.
3) Motion is propagation of oscillations in space and time, described by the electrodynamic laws.
4) The whole environment surrounding the man is penetrated by electromagnetic waves whose properties are caused by the ranges of their lengths and frequencies (Table 1).

3.4. Subject of Physics

As one of the fundamental components in natural sciences, physics studies both the elementary and most general properties of the material world. It is therefore rather difficult to strictly define physics scientifically as sets of various fields of knowledge of natural objects, phenomena, and processes (astrophysics, theoretical physics, the physics of solid state, gases and liquids, chemical physics, biological physics, radio physics and electronics, medical physics, etc.). It seems more expedient to consider establishment of most general and universal laws describing the motion of *substance* and *field* as a subject of physical research, on the basis of which it is possible to create a natural-scientific picture of the world. Physics is therefore the science studying most fundamental laws of the material world and taking a special place in the system of natural scientific cognition. Physical laws have universal character, i.e. the features of any material system, possessing its own specificity (chemical, geological, biological, etc.), are based on the laws of nature established by physics. Certainly, any more highly-ordered (biological) level of matter does not simply reduce to the physical level (reductionism). However, just in the development of physical representations, most constructive results in chemistry (quantum chemistry) and biology (molecular genetics) have been reached. Revealing fundamental laws of objective reality and possessing the necessary theoretical apparatus and empirical methods, physics has become the most developed field of modern scientific knowledge. The usage of the physical and mathematical apparatus and physical research methods in other sciences essentially raises their status in the cognition process dynamics. The regularities of the material world established by physics represent a system of statements (hypotheses, principles, theories, and laws) to give an idealized representation about physical objects. The set of these statements forms a physical picture of the world, i.e. an ideal model of nature corresponding to specific, historically caused stages of the development of natural scientific cognition and styles of natural scientific thinking. The physical picture of the world is based on our developing representations of substance and the forms of its manifestation, about physical force fields, about the specificity of spatial–temporal relations, about our understanding of the essence of motion and cause–effect relationships.

Classical and quantum mechanics, classical and quantum electrodynamics, equilibrium and non-equilibrium thermodynamics, statistical physics, the special and general theories of relativity, the quantum theory of field, etc. are the basic sections of theoretical physics. Essential development has been also gained by such leads of physical research as nuclear physics, the physics of elementary particles, the physics of plasma, solid-state physics, the physics of high and low temperatures and pressures, high-energy physics, etc. Fundamental developments have led to radical technico-technological discoveries in the sphere of applied physics, namely, nuclear power engineering, radar location, microelectronics, etc.

3.5. Principles of Classical Physics and the Limits of Its Applicability

The development history of physical sciences can be divided into several conditional stages, with *classical physics* (the 17–19th centuries) considered the most fruitful of which (preceding the appearance of modern physics). A short characteristic of some basic achievements of classical physics can be presented as follows.

Newton's classical mechanics was created: the notion of momentum was introduced; the law of preservation of the moment of momentum was formulated; the property of inertia (with mass as its measure) was characterized; the laws of mechanical motion and gravitational interaction were formulated. *Force* was established to be an action made upon a body with the purpose to change its state of rest or uniform rectilinear motion (force manifests itself in action only, and after the termination of this action does not remain in the body).

In the first half of the 19th century the basic law of nature was discovered, the law of energy conservation and transformation. *Energy* was shown to be a general quantitative measure of motion and interaction of all kinds of matter. According to various forms of matter motion, there also exist different kinds of energy, namely, mechanical, thermal, electrical, internal, chemical, nuclear, etc. The sense of the law of energy conservation and transformation is that in any isolated system the energy can pass from one kind to another one with its quantity remaining constant. The discovery of this law was basically contributed by the German doctor Mayer (1814–1878), the German physico-chemist German Helmholtz (1821–1894), and the English scientist Joule (1818–1889) who established the existence of the process of heat transformation into work and back, and, then, the processes of interconversion of various kinds of energy. This law, in essence, expressed the concept of non-annihilation of matter and motion.

The space and time relativity principle was offered (Galilei) and developed (Newton), which was one of the main preconditions for Einstein's creation of both special and general theories of relativity.

In the field of studying the processes of light electromagnetic wave propagation, several laws were established to describe the phenomena of light reflection and refraction, and Pierre Fermat's principle (the least time principle) to explain the straightforwardness of light propagation. The concept of this principle is that of all possible ways connecting two points, light chooses that one requiring the least time for its passage. The working principles of the lens were developed, allowing to determine its focal length and to explain the image magnification by means of a lens and the property of aberration associated with the distance of a light ray from the lens' axis. The laws of wave radiation of light described such phenomena as interference, diffraction, and dispersion, and the inversely proportional dependence of radiation energy on the squared distance. Light dispersion was shown to be caused by different factors of its refraction, and a special system of prisms can decompose light into seven colors (red, orange, yellow, green, blue, dark blue, and violet).

From the molecular-kinetic theory of gases it follows that heat is directly related with the vibrational motion of gas molecules, and its pressure results from impacts of the molecules with the vessel's walls. Hence, the pressure of a gas is proportional to the number of gas particles and the kinetic energy of their motion. The law associated with this theory and describing the behavior of an ideal gas ($PV = NkT$) established a strictly certain number of atoms N at identical temperature and pressure. Besides, if two gases have an identical temperature, then the average values of the kinetic energy of the molecules of these gases are also equal in the system of the center of mass. If the molecules move chaotically in three directions, then $(m\upsilon^2/2)_{av} = 3/2(kT)$; if in one direction only, then $\langle m\upsilon^2/2\rangle = 1/2(kT)$. An equal number of molecules are therefore contained in equal volumes of any gases at identical temperature and pressure. The molecular-kinetic theory of gases has allowed explaining such phenomena as diffusion, heat conductivity, dissolution, and calculating the average speeds of molecular motion and the average free molecule range (i.e. the value of the length of a rectilinear way passed by a molecule between two consecutive collisions). The molecular-kinetic theory of gases was logical completed by, first, van der Waals' law describing the behavior of a real gas:

$$(P + a/V^2)(V - b) = RT,$$

where a characterizes the intermolecular attraction forces, and b considers the molecular sizes, and, second, the formulation of the empirical gas laws by Boyle and Mariotte, Gay-Lussac, and Charles.

The bases of statistical mechanics were developed, which allowed obtaining, in particular, the quantitative dependences of the number of gas molecules (and density) on height:

$$n_h \sim \alpha \cdot e^{-mgh/kT},$$

where mgh is the potential energy of the molecules; and the speed distribution of molecules:

$$n_\upsilon \approx \beta \cdot e^{-m\upsilon^2/2kT}.$$

A theory of sound and its propagation was developed. Sound was shown to be vibrational motion of the elastic environment particles propagating in a wave mode, which includes the waves of both heard and unheard ranges. The man can hear sound of frequencies from 16 to 20,000 Hz. Sound with a frequency below and above the heard range is called infrasound and ultrasound, respectively.

An important characteristic of sound is its spectrum obtained by its decomposition into simple harmonious fluctuations and waves. The power characteristic of sound waves is the intensity of sound determined by the amplitude of sound pressure (or the vibrational speed of particles), the environment wave resistance, and the wave shape. It was also established that the phenomena of interference and diffraction usual for all types of waves were observed at propagation of sound waves and, when the sizes of obstacles (or inhomogeneities) exceeded the wavelength, sound propagation was described by the reflection and refraction laws, i.e. could be considered from the positions of geometrical acoustics.

Three laws (principles) of classical (equilibrium) thermodynamics were formulated. *The first law*: the change in the internal energy of a closed system is equal to the sum of the heat supplied to the system and the work done on it: $\Delta U = Q + \Delta W$. *The second law*: heat cannot spontaneously flow from a colder location to a hotter location. *The third law*: as a system approaches absolute zero (-273.15°C), all processes cease and the entropy of the system approaches a minimum value. The physical features of these laws will be discussed in more details in a special chapter devoted to the concepts of thermodynamics.

The known laws of electrostatics and classical electrodynamics (Coulomb, Ohm, Kirchhoff, Faraday, Maxwell, etc.), establishment of a relationship between mechanical and electric energies, revealing the causes of the occurrence of induced currents, and a relation between inductance and magnetic energy, establishment of the operation principles of waveguides were consequences of the electromagnetic concept.

The discovery of photoelectric effects by the Russian physicist Stoletov (1839–1896) was very important in fundamental and, especially, practical relation; further, it served as one of the starting points for the development of the concept of corpuscular-wave light dualism.

The Periodic Law discovered and formulated by the Russian chemist Dmitry Mendeleyev (1834–1907) had great importance for the confirmation of the concept of atomism (i.e. the actual existence of atoms) and formed a basis for disclosing the physical sense of the chemical element.

In 1895, the German physicist Wilhelm Conrad Röntgen (1845–1923) discovered radiation with high penetrability, i.e. the ability to pass through lightproof bodies, to ionize gases, etc. In a practical relation, this discovery has led to the design of technical devices allowing to explore the substance structure at an essentially new qualitative level (X-ray structural analysis), and to its usage in medicine to diagnose various diseases.

Studying the action of X-rays upon gases, J. Thomson (1856–1940) established a sharp increase of the electric conductivity of the gas, which remained even after the irradiation termination. This has led to a conclusion that electricity in gases is conducted by some charged particles formed under the influence of X-rays. A first elementary particle, electron, was so discovered (1896).

In the same 1896, the French physicist Becquerel (1852–1908) discovered the phenomenon of radioactive disintegration of chemical elements (blackening of a photographic plate having laid near uranium sulfate crystals), which (as well as the discovery of X-rays) played an exclusive role in the development of modern fundamental and applied physics and was the beginning of the development of nuclear physics.

At studying radioactive radiation by the English physicist Ernest Rutherford, radioactive atoms were found to release particles of two types, named as alpha particles (helium nuclei) and beta particles (electrons). Further, Maria and Pierre Curie succeeded to isolate a new element (radium) from uranium ores, which possessed a considerably higher radioactivity than uranium.

Analysis of the listed research results has allowed drawing some conclusions on the methodological principles of classical physics.

1) Physical systems as sets of material objects, phenomena, and processes exist objectively (i.e. irrespective of human consciousness); they are rather steady and located in a certain order in space and time.
2) The interaction of physical objects usually leads to no essential changes in their structure.
3) All physical systems (and their elements) existing in the real environment are related by cause-and-effect relationships which allow to unequivocally (i.e. precisely and simultaneously) determine the coordinates and behavior of the system (or its elements) through any time interval.
4) Experiment in which course the studied object (phenomenon, process) remains invariable is the basic criterion of reliability of the physical picture of the world. This means that any physical device used in an experiment must not affect the studied object directly to change its state.
5) Data about the properties of material objects and the mechanisms of phenomena and processes can be always expressed in a quantitative form (relationships, equations, formulae, etc.).
6) The majority of physical systems are closed, reversible (i.e. the direction of time has no special importance for them), and their properties, as a rule, obey linear laws.

At the same time, by the end of the 19th century it became clear that Newton's classical mechanics had a limited scope of applicability. First, it is inapplicable when the body motion speeds are comparable with that of light. In such cases it has been supplemented by *relativistic mechanics* designed on the basis of the statements of Einstein's special theory of relativity. Besides, from the mechanistic concept followed that the motion of bodies can be described by specifying their coordinates and the temporal dependence of velocity (i.e. the motion of bodies occurs along certain trajectories). However, experiments have shown such a description of motion being often inapplicable for bodies with a very small mass (e.g., for microparticles). This constitutes a second restriction of classical mechanics applicability. Then, it was shown that classical electrodynamics cannot explain many processes of light–substance interaction and the processes occurring within the atom (e.g., changes in the stability of atoms and their radiation spectra character). Finally, this led to the discovery of essentially new laws of nature, i.e. radical changes occurred in the system of natural scientific knowledge on the boundary between the 19–20th centuries, which affected its basic concepts. As a result of these scientific discoveries, the traditional representations of classical physics of the nuclear structure of substances were subjected to criticism. In connection with the electron discovery, the atom lost its status of a structurally indivisible element of matter. Newton's traditional representations about the Universe model according to which the world as a whole was considered as a stationary system were overcome. New experimental data of natural sciences, first, revealed the existence of not only the macrocosm but also a microworld; second, confirmed the representation of the relativity of scientific truth; third, showed matter to consists of not "indivisible primary element", i.e. the atom, but of a variety of kinds and forms of matter, phenomena, and types of interrelations.

Many new experimental facts could not be explained on the basis of the representations of classical physics associated with continuous changing of the energy of a physical system while its transition from one state to another one. It was necessary to find a way out from the arisen situation (this period is often named as the crisis in physics) which, on the one hand, would not contradict the bases of classical physics, and on the other hand, would involve essentially new ideas. In essence, the quantum concept was created in physics at this time. Quantum representations were first offered by the German physicist Max Planck (1858–1947) who put forward a hypothesis of the possible energy radiation by atoms of various substances not continuously but discretely, i.e. by certain portions named *quanta*. Planck's basic postulate consisted not only in discrete energy radiation but also in the energy of a quantum being proportional to the oscillator's (atom) frequency, i.e. the frequency of radiation: $E = h\nu$.

Just since that time, quantum physics and its sections (quantum mechanics, the quantum theory of the solid body, quantum optics and spectroscopy, the quantum theory of semiconductors, etc.) began their development. Such scientists as Max Planck, Albert Einstein (1879–1953), Erwin Schrödinger (1887–1961), Niels Bohr (1885–1962), Werner Heisenberg (1901–1976), Max Born (1882–1970), Enrico Fermi (1901–1954), Paul Dirac (1902–1984), Pauli (1900–1958), Lev Landau (1908–1968), Vitaly Ginsburg (1916–2009), Tamm (1895–1971), Frank (1908–1990), Basov (1922–2001), Prokhorov (1916–2002), Peter Kapitsa (1894–1984), Theodor Cherenkov (1904–1990), Abrikosov (b. 1928), Jores Alfyorov (b. 1930), et al. have contributed greatly to the creation and development of quantum representations.

Hence, in a most general view, the concepts of classical and quantum physics, respectively, can be formulated as follows:

1) when a physical system passes from one state to another one, its energy changes continuously;
2) when a physical system passes from one state to another one, its energy changes discretely.

3.6. Basic Types of Fundamental Interactions in the Macrocosm and Microworld

One of the major concepts of modern natural sciences implies that everything occurring in nature (the formation of substances, their transformations, the occurrence of force fields around macrobodies and microparticles, the influence of various kinds of energy on inanimate and animate nature) is caused by different types of interactions. Theoretical research and experimental facts give evidence of there being four types of physical interactions in the Universe, namely: gravitational, electromagnetic, strong, and weak. Investigating the mechanism specificity of these fundamental interactions is a task of modern physics.

The gravitational interaction is characteristic of all material objects irrespective of their nature. It consists in the mutual attraction of bodies and is described by the universal gravitation law. Falling of a body in the field of gravity, the motion of the planets of the Solar system, etc. are examples of the processes obeying this law. However, the determining role of gravitational interaction belongs to macroprocesses and interactions of macrobodies.

Gravitation features a relatively small intensity and a long-range action; i.e., despite its decrease with distance, gravitational interaction is observed at very long distances from the source. In particular, it plays the determining role in the megacosm, keeping planets on their orbits, stars within their galaxies, galaxies in their congestions, etc. It has no special significance in the processes in the microworld since in these cases it has to occur at very short distances and between insignificant masses.

The electromagnetic interaction is caused by the appearance of electric and magnetic fields and their interrelation. While the gravitational field is induced by all material objects, only charged particles are responsible for the electromagnetic field. It is necessary to always mean that the magnetic field arises only at motion of electric charges which, depending on their sign and direction of motion, either attract to or repulse from each other. Besides, as well as electric charges, the homonymic magnetic poles repulse while the heteronymic ones attract. But, unlike electric charges, magnetic poles exist only in pairs, namely, the northern and southern ones. Electromagnetic interaction may be exhibited at various distances from its source, i.e. at every level of matter (in the megacosm, macrocosm, and microworld). Just in this connection, many processes and properties of substances are determined by the electromagnetic interaction between particles (molecules, the atoms in a molecule, electrons, and atomic nuclei). Quantitatively, electromagnetic interaction is described by the known laws of electrostatics and electrodynamics and its intensity (as well as for gravitation) inversely proportional to the squared distance between the charged bodies.

The strong interaction is caused by nuclear forces, i.e. interaction of nucleons (basically, protons and neutrons) constituting the atom's nucleus, and is a source of huge energy. A characteristic example in this respect is the radiation of the Sun in which (and in other stars as well) the thermonuclear reactions caused by the strong interaction constantly proceed. The range of its action is determined by the sizes of the atomic nucleus, i.e. of an order of 10^{-13} cm. Therefore, in spite of the strong interaction essentially surpassing all the fundamental interactions by energy, it is almost not detected outside of the atomic nucleus. The name of such an interaction is due to that it is the most intensive and actually is responsible for the stability of atomic nuclei. The more strongly the nucleons in the nucleus interact, the higher the energy of their bonds is and the more stable the atomic nucleus (and, hence, the atom itself) is. At the same time, it is quite clear that with increase of the size of the atomic nucleus the energy between its nucleons should decrease. Just owing to this, the nuclei of the elements at the end of Mendeleyev's Periodic System of chemical elements are less stable and can even break up, as is shown in the phenomenon of radioactive disintegration of the atomic nuclei of these elements. It is very important that strong interaction is carried by *gluons*.

The weak interaction is less intensive in comparison with the strong one (though it also occurs with direct participation of elementary particles). Such an interaction proceeds much more slowly, is observed at insignificant distances from the source (shorter than 10^{-14} cm) and is short-range by duration. It was revealed at research of radioactive disintegration of chemical elements, and, in particular, at studying beta decay. It has been hypothesized (Pauli) that at β decay, one more elementary particle is released together with an electron, which has been named a neutrino (Fermi). Then, it was shown that neutrons left to themselves in an atomic nucleus during some time (usually few minutes) break up to three particles (a proton, an electron, and a neutrino). Therefore, the considered types of fundamental interactions can be arranged in the descending order by their relative intensity (energy): the strong, electromagnetic, weak, and gravitational ones. If the strong interaction is accepted as a unit, then the electromagnetic, weak, and gravitational ones will be lower by 10^2, 10^{10} and 10^{38} times, respectively.

In spite of the fact that the specified types of fundamental interactions essentially differ by their physical nature, they have some common principles.

One of such principles is due to that any type of interaction represents the influence of one object upon another one. The mechanism of such influence proceeding in space and time is caused by matter and energy exchange, which are motion characteristics. Therefore, it is possible to consider that there is always a feedback, which means that, first, there is no such motion which would not be accompanied by some interaction of matter elements; and, second, the interaction of material objects should lead to a certain change (including the character of movement). Hence, any interaction as well as motion is a form of matter existence.

Another general feature of all types of interactions is their universality concerning all kinds of matter. For gravitation, it is exhibited in attraction, for other interactions, both in attraction and repulsion between material objects. E.g., the mechanisms of many chemical reactions are caused by some electromagnetic interaction, in which the chemical bonds between atoms redistribute, which consists in spatial reorganization of the electronic shells in atoms. Even more specific example pointing to the generality of different kinds of interactions, can be the presence of versions of the physical form of motion: mechanical and thermal motions, changes of the state of aggregation of matter, crystallization processes, sound vibrations, nuclear and thermonuclear reactions. E.g., heat can be considered as a macroscopic form of motion, which is shown in collisions of atoms and molecules in gases, liquids, solid bodies, and plasma, and also as the interaction of radiation with substance. Temperature is a measure of the intensity of thermal motion, which is in direct proportional relation with the kinetic energy of particles or the energy of the absorbed radiation by the formula: $E = 3/2\ kT$.

3.7. Space and Time Relativity Concept. Conclusions of Einstein's Relativity Theory

In natural sciences, in particular, in physics, motion as one of the forms of matter existence is usually represented as changes in the state of any system in time and space. To quantitatively describe motion, special parameters are introduced, such as mass, force, velocity, acceleration, momentum (the quantity of movement), and some others. However, time and space are the primary and most general characteristics of motion, the representations about which changed at different stages of the development of physical sciences.

As early as in the 17th century, Newton considered that time could be absolute or relative. He understood the duration of some phenomenon or process as absolute (true) time. Newton considered the measure of the duration of some event as relative (or apparent) time, i.e. a more specific notion (a second, minute, hour, etc.). Then, it was shown that processes being long repeated with a certain degree, i.e. periodic processes (e.g., rotation of the Earth about its axis) are most convenient to clarify the physical sense of time. Further, it appeared that acyclic processes (e.g., radioactive decay of the atomic nuclei of some chemical elements, for which the half-life period is one of the basic characteristics) could be also used to measure time. At last, a concept of the modern natural sciences was developed on the basis of theoretical and experimental research, its sense was that time defined the change order of physical states of a system.

Developing the space and time relativity principle (which was first established by Galilei) in classical mechanics, Newton introduced the notion of "reference system". The reference system is a body in relation to which the location of any other moving body is defined at any instant. In physics, a special role is played by inertial reference systems, free physical bodies in which are either at rest or in uniform rectilinear movement relative to each other. For such systems, the relativity principle consists in that all mechanical processes proceed identically in all inertial systems. Practically, this means that two observers, who are in inertial systems, notice no changes in them. Such a statement is obvious for the mechanistic concept.

However, with the development of electromagnetic representations, a question arose of the generality of the space and time relativity principle. It was due to the incompatibility (at first sight) of the relativity principle with the postulate (and, later, the experimental proof) of the light velocity constancy in vacuum. Developing the bases of the special theory of relativity, Einstein has shown that: first, the specified contradiction is really apparent and caused by one of the concepts of classical mechanics according to which time and distance have absolute values, i.e. do not depend on the state of motion of the reference system; and, second, such a contradiction cannot basically exist, as the light velocity does not at all depend on whether the light source moves or is at rest. According to Einstein's theory of relativity, the notion of absolute time has no physical sense at all, as time (as a physical quantity) itself changes in different physical conditions, in particular, depends on the speed of motion of the reference system. Thus, the fundamental conclusion consists in that time always has relative character. Therefore, with a certain degree of accuracy it is possible to consider that only the phenomena (processes) identical in every respect proceed for the same time. At the same time, such a statement has relative character as well, since it cannot be checked because it is almost impossible to create absolutely identical conditions in nature at various times.

The representations of space were developing ambiguously as well. The founders of physiophilosophy defined the sense of space as though it simply caused some order of the existence of physical systems. In the Euclidian geometry, all mathematical objects were described without any dependence on space. Just from the statements of the Euclidian geometry followed the notion of absolute space which can exist and not change at the presence of bodies in it and at change of their properties. Developing his general theory of relativity, Einstein has united time and space into a uniform system (continuum), having shown that they cannot exist one without the other, and time represents a fourth dimension of space. From whence it follows that space, as well as time, is always relative, i.e. no absolute distance or length can exist.

Besides, the relativity principle applies not only to space and time but also to such an important physical quantity as mass. From the mechanistic concept (in particular, from Newton's second law) it follows that mass (as an inertia measure) is constant. However, according to the theory of relativity, mass must increase with increasing speed and, when speed is close to the light velocity, mass aspires to infinity. Just such a statement confirms the conclusion drawn earlier (in consideration of the electromagnetic concept) that no motion with a speed exceeding the light velocity is essentially possible.

At studying the problem of application of the relativity principle in different fields of natural sciences it is expedient to point to that the space—time system can have absolutely different senses. E.g., a perceptual (from the Latin *perceptus* – perception) “space—time” is known, being the space and time reflected in consciousness by means of the man’s sense organs. The conceptual (from the Latin *conceptus* – concept) “space—time” is specific logic models of space and time (e.g., phase, n-dimensional, etc. space and time). In physics, the notions of actual space and time (which, as a rule, are non-uniform and anisotropic) are commonly used. Nevertheless, the extent of space and the duration of time are general metric properties for all the specified systems. Besides, other “space—time” continua are also known in natural sciences, such as the geological, biological, psychological, and social “space—time”.

As an example concerning one of the components of such systems, it is possible to present the notion of biological time introduced by Vladimir Vernadsky. The average time of an organism’s life, the time of generation change, the time of doubling the numbers of a population, etc. are versions of biological time. It has been shown, e.g., that the time orientation of organisms is based on the so-called internal clocks whose presence is caused by rotation of the Earth about its axis, day and night changes, and, in general, the whole variety of biological processes. Internal clocks allow understanding many phenomena in animate nature, such as fish migration, bird flights, etc. Certain biorhythms are characteristic of the man as well. So, there exist physiological, emotional, and intellectual cycles which last, on the average, 23, 28, and 33 days, respectively. People are known to come under morning and evening groups (“owls and larks”) by their internal clocks. Psychological time is another notion. It is associated with features of the human perception of the duration of time and depends on the state of an organism, the character of his activity field, and the man’s age. Representatives of sociological sciences have introduced such notions as “social time” and “event space” which are caused by public conditions and relations.

In 1905, Einstein constructed his special theory of relativity from which a conclusion follows about the basic change of the spatial—temporal representations of classical physics. Lorentz's transformations gain a new sense; e.g. i) the same body has various length when moves with various speed relative to the system in which this length is measured; ii) the time interval during which this or that process proceeds is also different if measured by means of clocks moving with various speed. This means that the sizes of bodies and the duration of time have relative character since they depend on our choice of the reference system. The basic difference of the concept of the special theory of relativity from classical mechanics consists in that the description of motion with very high speeds is directly associated with the interaction of a physical process and its observer. Therefore, for the first time in the development history of physics, the dialectic character of cognition process was shown, exhibited in an interaction of the subject and object of cognition.

Further, Einstein has extended the space and time relativity principle to systems moving with acceleration, and to electromagnetic (e.g. optical) and other physical phenomena and has created the general theory of relativity (1916) which is the gravitation theory, or the theory of gravitational field. The substantive provisions and conclusions of Einstein’s special and general theories of relativity can be formulated as follows.

1) The basis of both theories is the postulate of the constancy of the light velocity in vacuum, and the invariance of physical laws at passing from one system to another; this means that all bodies or reference systems are equivalent whatever their state of motion is (inertial systems; systems moving with acceleration, etc.).

2) The interrelation between space and time has been established, which are necessary conditions of matter existence.
3) The properties of space depend on the properties of the bodies contained.
4) As the speed of motion of a physical system increases ($\upsilon \to c$), its mass increases, its length along the motion direction reduces, and time slows down:

$$m' = \frac{m}{\sqrt{1 - \frac{\upsilon^2}{c^2}}}; \qquad l' = l\sqrt{1 - \frac{\upsilon^2}{c^2}}; \qquad t' = t\sqrt{1 - \frac{\upsilon^2}{c^2}}.$$

5) The interrelation between mass and energy ($E = mc^2$) has been established, which means that the uniform law of energy and mass conservation acts in nature: every mass corresponds to its energy, every energy corresponds to its mass, i.e. energy release or absorption is always accompanied by mass loss or acquisition, respectively.
6) The invariancy principles in nature and the principle of space and time symmetry are associated with the relativity theory. Invariancy means the constancy of the properties of objects of nature and physical quantities at passing from one reference system to another one. The velocity of light in vacuum, the mass of an elementary particle, the charge of an atomic nucleus, etc. are, in particular, invariants. The space and time symmetry is most often shown in their uniformity. The uniformity of space means the invariance of physical properties of any closed system at its parallel translation in space. The uniformity of time means the invariancy of physical laws relative to the choice of a reference mark of time (e.g., at free falling of a body in the field of gravitation its speed and way depend only on the initial speed and the duration of falling, but do not depend on when the body started to fall).

The major experimental acknowledgements of the justice of the general theory of relativity are:

- the deviation of a light ray from its rectilinear propagation at passage near a body with a big mass, i.e. in the field of gravitation;
- the displacement of the elliptic orbit plane (the motion trajectory) of a planet round the Sun whose value increases with reduction of the distance to the Sun, i.e. with the gravity growth;
- the shift of the spectral lines of star radiation towards red color (the so-called red shift, or Einstein's effect), i.e. reduction of the frequency of light radiation of stars under the influence of a gravity field.

In summary, it is necessary to pay attention to one important circumstance. The space and time relativity concept by no means contradicts the principle of the relativity and limitation of scientific cognition. This is due to that both Einstein's special and general theories of relativity apply to the processes occurring in the objective world while the relativity of knowledge is often caused by subjective representations of objectively occurring natural processes.

3.8. Symmetry and Asymmetry Concepts in Nature

As follows from the previous section, the symmetry principle in nature is one of the consequences from Einstein's relativity theory. However, the notion of symmetry in natural sciences and, in particular, in physics has a deeper sense, first of all, in connection with the action of the laws of preservation of various quantities which cause the stability of the environment. These are: 1) the law of conservation of energy and mass which is associated with the constancy of these quantities at passing of energy from one form in another one; 2) the laws of preservation of momentum and the moment of momentum reflect the generality of translational and rotary movements, respectively; 3) the law of preservation of electric charge is caused by the presence of electromagnetic interaction; 4) Galilei's (in classical mechanics) and Lorentz' (in classical electrodynamics) mathematical transformations; 5) the conclusions from Einstein's special theory of relativity, which point to the invariance of some physical quantities at passing from one reference system to another one. Therefore, the notion of symmetry (from the Greek *symmetria* — harmony) generally means the invariancy of the structure and certain properties of a material object at its transfer from one reference system to another one.

Two forms of symmetry are characteristic of physical systems. *Geometrical symmetry* reflects the properties of space and time and manifests itself in its three versions, namely: *mirror symmetry* (having a line, a plane, or a time instant between two parts of a single whole absolutely identical to each other), *rotary symmetry* (the presence of some center around which occurs repeated turning of the same structural element), *translational symmetry* (multiple repetition of the same structural element in space and time). *Dynamic symmetries* express internal properties of objects and phenomena and are caused by fundamental physical interactions (their characteristic will be given below). *Calibration symmetry* is one of the most often met versions of dynamic symmetry, among its consequence is, e.g., the law of electric charge preservation.

Numerous researches point to that the symmetry concept has very great importance in quantum physics whose representations underlie the modern natural scientific picture of the world. As a symmetry example in quantum mechanics it is possible to present the existence of antiparticles (a positron, an antiproton, an antineutron).

At the same time, it is necessary to mean that, along with symmetry in nature, the asymmetry principle is also realized, i.e. heterogeneity, disproportion, disorder, etc. E.g., in mechanics, the states of rest and uniform rectilinear movement are symmetric, while those of rest and accelerated motion are asymmetric. The unity of symmetry and asymmetry in nature is caused by that these concepts always precede one another, and, finally, this unity manifests itself in the development of the process of cognition of the real world. This can be exemplified by the coexistence of opposite magnetic poles, the corpuscular-wave dualism of light and microparticles (which will be considered below as well), and, in general, the interrelation between the laws of physics which describe various physical properties and phenomena caused by both symmetry and asymmetry. Moreover, each law in relation to some specific conditions and to their changes is asymmetric, namely: the laws of classical mechanics is asymmetric relative to Lorentz' transformations, the law of speed distribution of gas molecules establishes a relationship between the average speed and the minimum and maximum values, the second law of thermodynamics (entropy increasing as a result of transformation of any kinds of energy to the thermal one) is asymmetric relative to possible transfer of various energy kinds to each other, etc.

Thus, the presence of objective interrelations between symmetry, asymmetry, and the universal laws of physics allows understanding their role in the course of nature cognition.

3.9. Quantum Concepts in Modern Physics

3.9.1. *Atomic* Structure Theory

While the representations of atoms as of the smallest indivisible particles arose in the antique times, their existence was proven in the 18–early 19th century only, basically, in the works by Lomonosov (1711–1765), Lavoisier (1743–1794), and Dalton (1788–1844). A special role in this problem was played by Dmitry Mendeleyev (1834–1907) who developed the Periodic System of chemical elements in 1869 and so raised the question of the uniform nature of atoms. A more strict scientifically founded arrangement of elements became possible after the discovery of the electron by J. Thomson (1856–1940) who offered the first model of the atom, correctly treating it as neutral (though the existence of an atomic nucleus was not proven yet). Then, have been defined the mass ($9.1 \cdot 10^{-28}$ g) and charge ($1.6 \cdot 10^{-19}$ C) of the electron were evaluated. The positively charged part of the atom, i.e. its nucleus, was discovered by Rutherford (1871–1937) in 1911, who gave a dynamic character to Thomson's static model.

The scheme of the atomic structure offered by Rutherford (it is often named the planetary model) consists in the following. There is a positively charged nucleus at the center of the atom in which almost the whole mass of this atom (as the mass of the electrons is insignificantly small) is concentrated. Electrons rotate round the nucleus on different orbits. The centrifugal force arising while their rotation is counterbalanced by the force of attraction between the nucleus and the electrons, in which connection the electrons are located (naturally, on each orbit) at certain (constant) distances from the nucleus:

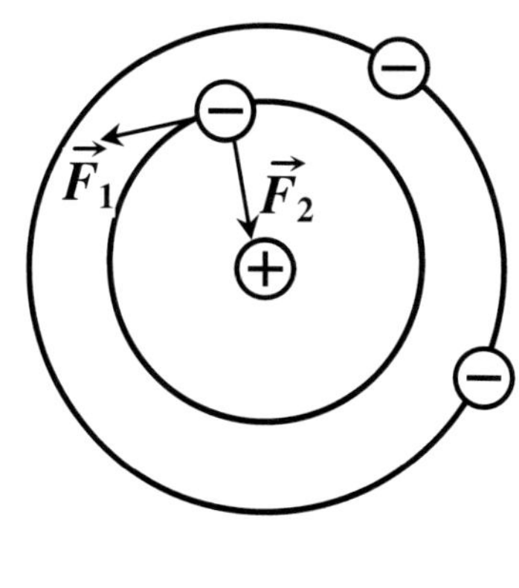

$$|\vec{F}_1| = |\vec{F}_2|$$

The further development of the atomic structure theory (Rutherford's model) was associated with the explanation of one of the new experimental facts, namely, the presence of line spectra obtained at decomposition of the light radiated by atoms in vapors and gases:

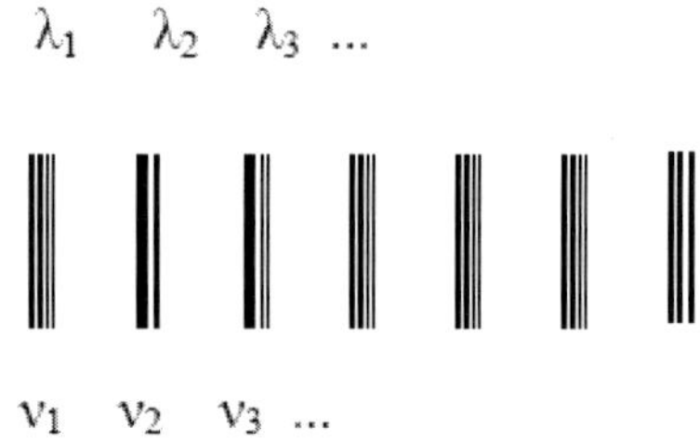

Rutherford put forward a hypothesis according to which the line character of these spectra was caused by oscillation of the electrons in atoms, which are equivalent to their rotations round the nuclei. It was then hypothesized that a rotating electron emitted light with a certain wavelength depending on the frequency of its rotation on its orbit (i.e. on the frequency of apparent oscillations), and, thus, each line of such a light spectrum has its "own" wavelength (λ_1, λ_2, λ_3, etc.) and the frequency of oscillation (ν_1, ν_2, ν_3, etc.). However, such representations by Rutherford led to a basic contradiction. The matter is that an electron emitting light loses a part of its energy, so the balance between it and the nucleus must be broken. For balance restoration, the electron must necessarily come nearer to the nucleus. Finally, having exhausted its whole energy, the electron must fall to the nucleus and light radiation will stop. From whence it is possible to make two conclusions, namely: first, if the electron actually radiates light continuously, the spectrum must be continuous rather than line; second, falling of the electron on the nucleus means atom destruction, i.e. the termination of its existence (in essence, destruction of substance), which actually is not observed.

The recovery from such a situation was found in an essentially new approach, namely, in the quantum theory of the atomic structure, which was offered by Niels Bohr (1885–1962) in 1913. Two postulates underlie his concept. The first postulate consists of three statements:

- the atom can exist in stationary (i.e. not changing with time) states, in which it radiates no energy;
- these stationary states of the atom have stationary (fixed by energy value) orbits, which the electrons move on);
- the electrons' motion on these orbits is accompanied by no energy radiation.

The second postulate concerns the possibility of the electron's transition from one orbit to another one, which, in turn, leads to transition of the atom itself from one energy state to another one. Therefore, the essence of the second postulate is in that the transition of the electron from one stationary orbit to another one is accompanied by either radiation or absorption of one quantum of light (a photon), with the energy equal to the energy difference of the corresponding stationary states of the atom:

$$E = E_n - E_m \text{ or } E = E_m - E_n .$$

This means that when $E_n \to E_m$ a photon is emitted, i.e. the atom passes from its state with a higher energy into that with a lower energy; and, on the contrary, when $E_m \to E_n$ a photon is absorbed, i.e. the atom passes from its state with a lower energy into that with a higher energy. Whence it follows that the atom radiation is characterized by a set of possible frequencies of quantum transitions

$$\nu_i = |(E_n - E_m) / h|.$$

Just such a set of frequencies causes the line (discrete) spectrum of atom radiation.

3.9.2. Corpuscular-wave Dualism Theory of Light and Microparticles

The *wave representations of light nature* arose as early as in the 17th century when a hypothesis was stated of light being oscillation propagating in a special environment ("ether") which fills the whole world space. In a general sense, the *wave* as a physical notion is excitation (or indignation) which propagates in space with a finite speed, and with energy transfer but without mass (substance) transfer. The key parameters of a wave are its length (λ), speed ($\acute{\upsilon}$), frequency (ν), and the oscillation period (T) (e.g., the light velocity is $c = \lambda\nu$).

The wave theory of light is underlain by Huygens—Fresnel's principle according to which every point having been reached by the excitation becomes a center of secondary waves and transfers them extensively to next points. The phenomena of interference, diffraction, and dispersion can be considered as experimental evidence of the wave properties of light. Light interference consists in strengthening or easing of oscillations (light) on mutual imposing of two and more waves. Diffraction is the deviation of a light ray from its rectilinear direction on its meeting with any obstacle. Light dispersion (or decomposition) was discovered by Newton who established the dependence of the light propagation speed on its wavelength.

The *representations of the quantum properties of light* appeared since the discovery of the photoelectric effect, i.e. the phenomenon of electron emission from the atoms of substances under the influence of an incident light. The occurrence of free electrons in substance leads to the appearance of an electric current. The quantitative photoeffect laws were established by Stoletov (1839–1896):

- the magnitude of the arising electric current is directly proportional to the intensity of the light radiation falling onto the body's surface;
- the maximum kinetic energy of photoelectrons does not depend on the intensity of light radiation but increases with the frequency of light;
- at a frequency of light below some frequency existing for the given substance, no photoeffect is observed (the so-called red limit).

It was not possible to explain the specified regularities on the basis of the wave theory of light. The basic concept of the quantum theory offered by Einstein in 1905 consists in that light propagation as well as radiation occurs as certain portions, i.e. a flow of light quanta (photons) whose energy is determined by Planck's formula:

$$E = h\nu,$$

and the mass of a photon is related with the wave characteristics of light:

$$m = E / c^2 = h\nu / c\lambda\nu = h / c\lambda.$$

So, it is possible to draw the following fundamental conclusions:

- light has dual, corpuscular-wave nature;
- light represents the unity of opposite properties, corpuscular (quantum) and wave (electromagnetic) ones;
- light represents the unity of continuity and discreteness.

Developing the corpuscular-wave dualism of light and meaning such a property of nature as symmetry, Lois de Broglie (1892–1987) put forward a hypothesis about the universality of the corpuscular-wave dualism in 1923. He considered that electrons and any other particles of matter as well as photons possessed such a dualism of their properties. According to his concept, the corpuscular characteristics (energy, mass, momentum) are associated with every microobject on the one hand, and, on the other hand, the wave ones (the frequency of oscillation and wavelength). The quantitative relationships are identical for all particles. The main conclusion consists in that there is a potential possibility for every microobject to exhibit itself, depending on external conditions, either as a particle or as a wave.

In a practical relation, the theories based on the concept of corpuscular-wave dualism have led to the creation of such new physical techniques of exploring matter as crystallography, electronography (electronic microscopes), neutronography (tomographs).

3.9.3. Some Principles of Quantum Physics

Principle of Uncertainty

According to classical mechanics, any body moves along a certain trajectory and at every instant it is possible to precisely find its coordinates and momentum. Moreover, using the known equations of dynamics, classical mechanics can make predictions in the global plan as well: the presence of new planets, celestial motion laws, and features of various phenomena occurring in the Galaxy. Unlike it, the concepts of quantum mechanics have a probability character. The corpuscular-wave nature of microparticles is another basic difference of quantum physics from classical one. This does not mean that microparticles can possess both wave and corpuscular properties simultaneously. According to the quantum representations, the notion of "the wavelength at a given point" has no physical sense. Hence it is possible to draw a conclusion that an electron cannot be considered as a material particle precisely localized in time and space. The modern physics can provide either the exact position of a particle in space with its uncertainty in time, or, on the contrary, the exact time with some uncertainty in space. Such a statement is in full conformity with the concept of the continuous-discrete properties of microparticles, which allows defining their momentum through the wavelength:

$$P = mc = h / \lambda.$$

Just from the relation between the mass and wavelength it follows that a microparticle with a certain value of momentum has an uncertain coordinate, and *vice versa*. And in general, the more precisely is defined one of the quantities related by the equations of quantum mechanics, the less exactly appears the value of another quantity. This principle first was offered in a quantitative form by Werner Heisenberg (1901–1971) in 1927 as one of the laws of quantum physics and was named the uncertainty ratio:

$$\Delta x \Delta p \geq h,$$

where x is the coordinate; p the momentum; Δ is the increment of the quantity.

This means that the product of the uncertainties (or discrepancies) of the coordinate and momentum cannot be less than Planck's constant.

Complementarity Principle

The uncertainty principle is very closely related with another basic statement of quantum mechanics, namely, the complementarity principle. In essence, it is the same principle, only concerning a fewer numbers of specific quantities related with each other. The complementarity principle was offered by Niels Bohr in 1927 to resolve the theoretical difficulties arisen in the course of interpretation of the physical nature of microparticles. According to this principle, reliable obtaining experimental data of some physical quantities characterizing some microobject (an elementary particle, an atom, a molecule), is necessarily accompanied by loss of reliable information about other quantities complementary to the first ones. Such complementary (to each other) quantities usually are the coordinate of a particle and its velocity (momentum), the value and direction of the momentum, energy and time, kinetic and potential energy, etc.

Generally, the essence of the complementarity principle consists in that for reproduction of the integrity of a phenomenon at a certain stage of its cognition, application of mutually exclusive and mutually limiting classes of notions and representations is necessary. Just their set provides the relative completeness of information. It is a question of the basic necessity of the involvement of various theoretical approaches to describe the interaction between the subject and object of cognition, and of the impossibility to reach the level of absolute (true and finished) knowledge. As it has already been told, microparticles can show both corpuscular and wave properties, and this apparent contradiction is overcome in quantum mechanics on the basis of the representations of corpuscular-wave dualism. This means that both aspects of the electron (corpuscular and wave) mutually supplement each other. Only joint application of both representations gives a more complete description of the physical essence of light and microparticles. The use of different classes of devices with whose help it is possible to maximize the accuracy of measurement by separate parameters is thus essentially inevitable. This is reached by refusal from unequivocal (exact and simultaneous) measurement of other characteristics of the object under study since any measurement, influencing the object, anyhow influences other aspects of its behavior. In this connection, it is necessary to use both wave and corpuscular theories and to apply devices of absolutely different classes for cognition of the properties of an electron and other microobjects.

In spite of the fact that the complementarity principle has been formulated to solve specific quantum-mechanical problems, its application leaves far beyond direct physical knowledge. In particular, this principle acts as one of the factors of overcoming of the contradictions between physics and biology. The basis for application of the complementarity principle within the framework of biological systems is the integrity of living organisms which is combined with their divisibility. The importance of this principle in biology increases in connection with the development of the concepts of the structural organization of live matter. On the one hand, the problem of levels in biology is related with the objective structured nature of living organisms, and, on the other hand, the profound cognition of biological objects should not be associated with loss of their integrity. Hence, the use of the methods of physics (and mathematics and chemistry) for studying life is complementary in the course of comprehension of the essence of biological systems.

Conformity Principle

According to this principle, any replacement of one natural-scientific theory by another one causes not only their development but also the continuity between them. The new theory not simply denies the previous representations but includes them in a certain form, thereby establishing the limits of the old theories and planning prospects of the new concepts. The conformity principle was also formulated by Niels Bohr in 1913, having shown that the laws of quantum mechanics pass into the laws of classical mechanics when the quantum of action is neglected. On the other hand, the laws of the theory of relativity pass into the laws of classical mechanics when the velocity of light is considered infinite. Just in this the interrelation between the new and old representations is shown. Therefore, physical theories replace each other not chaotically, completely denying one another, but organically join into the infinite process of cognition.

The conformity principle formulated for physical reality has its applications in other spheres of scientific knowledge as well. E.g., molecular biology is not only a new stage in the development of the whole complex of sciences about animate nature (and a new discipline of the biological cycle) but includes some concepts associated with traditional researches in biology (for example, with the concept of natural selection). Thereby, the sciences of living substance, leaving on a new level of cognition, do not reject the traditional representations but reveal their interrelation.

3.9.4. Superfluidity and Superconductivity Phenomena

The *superfluidity phenomenon* is connected with obtaining liquid helium and its properties which cannot be explained on the basis of the representations of classical physics. Before the discovery of this phenomenon, it was known that near absolute zero (–273.16ºC)` any substance under any pressure must be in a solid state. This is caused by that, according to the concepts of classical physics, the kinetic energy of atoms is equal to zero at absolute zero, i.e. all atoms are motionless. Such an arrangement of atoms must be ordered and represent some spatial lattice. Hence also follows that any substance must be crystal at absolute zero. However, it has appeared that there is an exception, namely: helium after its liquefaction remains liquid at all temperatures, even rather near absolute zero. Two liquid helia are known. If helium passes into a liquid state at 4.2 K (or –268.8°C) it is helium-1. At 2.2 K (–270.8°C) it, remaining liquid, undergoes another transformation (a second-order phase transition); it is helium-2.

If a first-order phase transition represents transition of substance from one aggregate state to another one (e.g., liquid—gas, solid—liquid), at a second-order phase transition, one crystal modification passes into another one. It is experimentally established that the aggregate states of both crystal modifications coincide, i.e. there is no that jump which is observed at any first-order phase transition. This, in turn, leads to the absence of jump in the quantities characterizing the thermal state of the body (volume, internal energy, and other thermodynamic functions). Just therefore such a transition is accompanied by no heat release or absorption. At the same time, at the transition point (from one crystal modification to another one) there occurs a jump-like change of the character of the temperature dependences of physical quantities, namely: the thermal expansion factor $(dV/dT)_P$, the body's thermal capacity $(dQ/dT)_P$, the compression factor of the body $(dV/dP)_T$, etc. Besides, it is experimentally proven that liquid helium possesses various properties. In particular, one of the properties of helium-2 is a very high rate of heat transfer inside it, caused by a rather high factor of heat conductivity. However, the most main property of liquid helium-2 is superfluidity consisting in a sharp decrease in the value of viscosity (almost down to zero). It is pointed, for example, by such an experimental fact as the process of spontaneous overflowing of liquid helium from one vessel into another one, which are separated with a partition (and the rate of overflowing is not less than 10 cm/s):

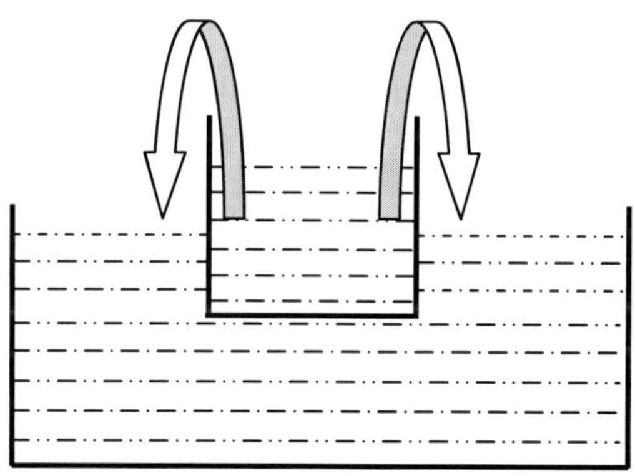

It is possible to explain the superfluidity phenomenon by the assumption that liquid helium-2 represents a so-called quantum liquid, i.e. a liquid whose properties are caused by quantum effects in the behavior of its particles. This concept was developed by Kapitsa, Landau, and Ginsburg. Generally, it is known that at a fall of temperature the role of quantum effects always increases, but many usual liquids harden earlier than these quantum effects start to manifest themselves. However, in most cases, considerable quantum effects are shown at very low temperatures when de Broglie's wave of the particles, corresponding to their thermal movement, becomes comparable with the distances between them. Just at this instant a so-called quantum degeneration of a liquid occurs. For crystals, the quantum effects are changes in the thermodynamic and kinetic properties of a crystal when at least one of its geometrical dimensions becomes comparable with the length of de Broglie's wave of electrons. Hence, according to the concept of corpuscular-wave dualism of microparticles, it is possible to consider that if a particle has an energy $E = h\nu$ and momentum $p = mc$, a wave with a frequency $\nu = E/h$ and length $\lambda = h/p$ ($h = 6.7\times10^{-27}$ erg·s) is associated with it. Such waves are called de Broglie's. Generally, $\lambda = h/mv$, i.e. the higher m and v, the lower λ is. Therefore, a particle with a 1 g mass and moving with a 1 km/s speed must correspond to the length of de Broglie's wave of 10^{-19} nm, which lies outside any area accessible to observation.

The helium isotopes, liquid ^{4}He and ^{3}He within 2–4 K, are the most known quantum liquids. The classification of a quantum liquid depends on the value of its spin, namely: the liquids consisting of particles with an integer spin (i.e. bosons) are named Boze-liquids (^{4}He), while those with a half-integer spin (i.e. fermions) are Fermi-liquids (^{3}He).

The *superconductivity phenomenon* was discovered by Camerling-Onnes in 1911 at studying the temperature dependence of the electric resistance of mercury. It was shown that at T = 4 K its resistance jumped to zero. Then, it appeared that this phenomenon was shown for many chemical elements, their compounds, and even alloys at cooling below a certain temperature T_C (characteristic for every material). At transition from the normal state to the superconducting one, the optical properties sharply differ from the properties at usual temperatures (including the electric resistance to direct current is almost equal to zero). Then, researches of this phenomenon showed that the normal state can be restored by carrying an enough strong current to the sample or its placing into a strong magnetic field. The values of the current strength and the magnetic field intensity are also characteristic for every material.

As the superfluidity and superconductivity phenomena occur at the same (low) temperatures, Landau et al. have decided to offer a uniform explanation to both phenomena. First, they assumed that superconductivity was caused by the superfluidity of a quantum liquid, and then derived equations describing the thermodynamics and electromagnetic properties of superconductors. In further research (Frelich and Bardin, 1950) the dependence of the critical temperature of transition into the superconducting condition (T_C) on the metal nature was established. This gives evidence of superconductivity being influenced by some features of the crystal lattice. A conclusion has been drawn on the possibility of a specific attraction between electrons, which is much stronger than Coulomb's repulsion among them. Then, in 1956, Cooper experimentally proved that under these temperature conditions there really appear specific bonds between electrons (so-called Cooper's couples), which was a basis for development of the theory of superconductivity (Bardin, Cooper, Schriffer, 1957). The electron coupling phenomenon has been shown to be due to the presence of a certain modification of the crystal lattice and its influence on the electrons. This is caused by the crystal lattice being in a vibratory state (zero vibrations) even at a temperature close to absolute zero. The electrostatic interaction of an electron with the lattice ions changes the character of these vibrations, which leads to the occurrence of an additional attraction force acting upon another electron, to coupling of these electrons, and, finally, to a decrease in the electric resistance.

In a practical relation, studying the phenomena of superfluidity and superconductivity, which have, in essence, the same physical nature, has allowed synthesizing new materials possessing unique properties.

3.9.5. Elementary Particles: Their Properties and Characteristics

Beginning from ca. the 1930s, the history of modern physics is connected with the development of the atomic structure theory and nuclear reactions, and nuclear physics appeared as a result. On the one hand, this lead of physics studies the structure, properties, and interconversions of atomic nuclei. On the other hand, it directly includes the physics of elementary particles and atomic power engineering. The notion "elementary particles" is associated with the establishment of the discrete character of the substance structure at a microscopic level and is used to name a large group of smallest particles possessing quantum properties and being neither atoms nor their nuclei (except for the proton). To put it briefly, elementary particles are really indivisible particles which matter consists of.

The establishment of molecules being built of atoms has allowed, for the first time, describing all known substances as combinations of a finite number of their structural components. The discovery of electrons and atomic nuclei, revealing the composite nature of nuclei appeared built from protons and neutrons, have essentially reduced the number of the discrete elements forming the properties of substance. Then, however, it has appeared that protons and neutrons may have a more complex structure. Except the electron, proton, and neutron, some other particles belong to the elementary ones, namely, the photon, pions, muons, neutrinos of three types (the electron neutrino, muon neutrino, and tau neutrino), so-called strange particles (kaons and hyperons), various resonances, psi-particles, "fascinated" and upsilon particles, heavy leptons – above 350 sorts in total.

The discovery of elementary particles was prepared by research of the optical spectra of atoms, studying electric phenomena in liquids and gases, the discovery of such phenomena as the photoelectric effect, X-rays, radioactive decomposition of atoms, which give evidence of a complex structure of matter. Historically, the electron was the first discovered elementary particle, when in 1896 Sir Joseph John Thomson established that cathodic rays were formed by a flow of smallest particles, carriers of a negative elementary electric charge in atoms. In 1911, Ernest Rutherford, passing alpha particles from a natural radioactive source through a thin foil of various substances, found that the positive charge in atoms was concentrated in their nuclei. In 1919, he found particles with a unity positive charge among the particles dislodged from atomic nuclei, named protons (their mass exceeds that of a electron by 1,840 times). Another particle, a part of a nucleus (the neutron) was discovered by Chadwick when studying the interaction of alpha particles with beryllium in 1932. The neutron has its mass close to that of a proton but possesses no electric charge.

In more details, the discovery history of elementary particles and the development of the knowledge of their properties look as follows. Developing Planck's idea of the possibility of energy radiation by atoms of various substances in a discrete manner, Albert Einstein has shown that light as electromagnetic radiation is a flow of separate quanta (photons). The existence of photons has been experimentally proven by Millikan and Compton. The discovery of the neutrino is associated with Pauli's hypothesis, allowing eliminating the difficulty with the law of energy conservation in the processes of beta disintegration of radioactive nuclei. Experimentally, the existence of the neutrino was confirmed in the late 1940s–early 1950s in the works by Fermi, Reines, and Cowan. In general, studying elementary particles is most closely connected with space ray research. E.g. in 1932, Anderson found a positron, the first antiparticle possessing an electron's mass but a positive electric charge. The positron existence directly followed from the relativistic theory of the electron developed by Paul Dirac. In 1936, the American physicists Anderson and Neddermeyer found muons in space rays (having both signs of electric charge), particles with a mass equal to ca. 200 electron ones but rather close to the electron and positron by properties. In 1947, in space rays as well, Powell's group discovered Yukawa-predicted pi mesons, or pions (π^+ and π^-) with a mass of 274 electronic ones, which play an important role in the interaction of protons and neutrons in nuclei. The 1950s were marked by the discovery of a big group of particles with unusual properties, named “strange” ones (K^+ and K^- mesons, etc.). The accelerators of elementary particles turned to be the basic tool for their research and the energy of the particles accelerated in these accelerators were ten and hundred billions eV in the 1970s. The application of elementary particle accelerators has allowed proving the existence of heavy antiparticles (antiproton, antineutron, and antisigmahyperons). In the 1960s, a great number of extremely unstable particles were discovered, which received the name of “resonances”, the mass of the majority of whom exceeding that of a proton. In 1962, the existence of two different neutrinos, the electron and muon ones, was proven. In 1974, rather massive (3–4 proton masses) and, at the same time, rather steady psi-particles were found with their life period unusually long for the resonances. In 1975, first data on the existence of heavy analogs of the electron and muon (a heavy lepton) were obtained. In 1977, information of gamma particles with their mass of ca. ten protons was obtained. Particles named bosons were theoretically predicted in the 1970s and experimentally discovered in 1983, being the carriers of the weak interaction between electrons, protons, neutrons, and neutrino. Besides, they have appeared to be neutral or charged. Therefore, for the years passed after the discovery of the electron, a huge number of various microparticles of matter have been revealed.

Exploring the internal structure of matter and the properties of elementary particles was accompanied by radical revision of many early laws of classical mechanics and electrodynamics. The concepts of the special and general theories of relativity, quantum mechanics, and the quantum theory of field form the basis for an essentially new description of the microworld phenomena. Elementary particles participate in all the types of fundamental interactions and so can be divided into several groups.

Leptons. The electron and two kinds of muons (they do not participate in the strong interaction only) and the neutrino which participates in neither strong nor electromagnetic interactions are most known of them. There are three types of neutrino, namely: the electron, muonc, and tau neutrinos, and each lepton has its own antiparticle, i.e. the total number of leptons is twelve.

Hadrons. They participate in all the types of fundamental interactions, except the gravitational one, can be neutral or charged. Protons and neutrons are most known of them, and the others (their number reaches hundreds) are extremely unstable and represent two kinds, namely: baryons (heavy hyperons and baryonic resonances) and mesons (mesic resonances). In 1963, it was established that the hadrons are combinations of two or three structural elements named quarks and possessing the smallest mass in comparison with all the known elementary particles. Gluons are the carriers of the interaction between these quarks (the strong interaction) to link them in couples or triples. Quarks have a fractional electric charge which is either –1/3, or +2/3 of the electron charge. The combination of two or three quarks can have the total charge equal to zero or unit.

Photons. They, being neutral particles and having zero mass at rest, transfer only the electromagnetic interaction between charged particles.

Graviton is a so-called hypothetical elementary particle, for which the gravitational interaction is characteristic (till now its existence has not been confirmed experimentally, as gravitation is least intense of all the known fundamental interactions and does not accept participation in quantum processes).

Research of the participation of elementary particles in different types of fundamental interactions has allowed making two general and very important conclusions:

1) all the interactions represent the processes of originating and destruction of microparticles;
2) all the processes of originating and destruction of particles occur in neither space nor time but instantly and are localized at a certain point.

Elementary particles can be characterized by other properties as well. First, by their life time, the elementary particles are divided into stable (the electron, proton, photon, neutrino), quasi-stable (the neutron, mesons), and the resonances ($10^{-24} - 10^{22}$ s). Second, each particle possesses its own mass at rest measured in MeV. The range of mass values of various particles is rather wide: from zero (the photon and neutrino) to 3,000 MeV and higher. The initial classification of particles has been based on the values of their masses. But this quantity (as well as the atomic mass in Mendeleyev's Periodic System of chemical elements) has appeared to be necessary but insufficient. Third, one of the major characteristics is the own moment of momentum of an elementary particle, which is a rotating whipping top. This quantity is called *spin* (usually designated as *s*):

$$s = m \cdot \upsilon \cdot r,$$

where r is the radius; m the mass; υ the rotation speed of the particle (top).

According to the statements of quantum mechanics, the spin can accept only discrete values, integer or half-integer but multiple of Planck's constant. E.g., for the electron and proton $s = 1/2$, for the photon $s = 1$ (though it is not absolutely exact as its mass at rest is zero). Therefore, depending on the spin value, some groups of elementary particles are also divided into subgroups. In particular, hadrons are divided into baryons (with half-integer spins – protons, electrons, neutrons) and mesons (with integer spins).

It has been also established that a particle's spin unequivocally determines the statistics type: the particles with half-integer spin (the leptons, baryons, and baryonic resonances) obey Fermi–Dirac's statistics, while those with integer spin (the photon, mesons, and mesic resonances) obey Bose–Einstein's one. Fourth, every particle is characterized by its electric charge which can take on integer values only (the charge of an electron is accepted as -1). Now, a plenty of neutral particles and particles with their charge equal to unit by magnitude are known.

To characterize elementary particles, the lepton number L and the baryon number V have been introduced. By definition, for the leptons $L = +1$, $B = 0$; for the baryons $L = 0$, $V = +1$; for the mesons and photon $L = V = 0$. In all reactions with the participation of elementary particles these quantum numbers remain.

And, at last, as it has already been noted, besides usual particles, antiparticles are known to have the same mass, lifetime, and the spin value but to differ by the charge sign (electron–positron, proton–antiproton). A general feature of particles and antiparticles is their ability of annihilation at collision, i.e. to transformation into other particles with preservation of the total energy, momentum, electric charge, and some other characteristics. An example is the appearance of two photons at interaction between an electron and a positron.

3.9.6. Features of the Atomic Nucleus Structure and Nuclear Reactions

The nucleus of an atom is its central part having a positive charge, in which almost the whole mass of the atom (the sum of the masses of nucleons, i.e. protons and neutrons) is concentrated. The proton possesses the mass at rest of 1.6726×10^{-27} kg and the positive electric charge equal (by magnitude) to that of an electron (1.6×10^{-19} C). The mass at rest of a neutron is 1.6749×10^{-27} kg. The charge of the nucleus of an atom is calculated as the product of the serial number of a chemical element in Mendeleyev's Periodic System (equal to the number of protons in the nucleus) on the proton charge value. The total number of nucleons in the nucleus is the mass number. The chemical elements whose atomic nuclei have an identical number of protons but different mass numbers are called isotopes. If the mass number is the same and the number of protons varies, such elements are called isobars. The size of the nucleus is characterized by its radius which has conditional sense due to the uncertainty of the borders of the nucleus. Nucleons possess their own magnetic moment measured in the so-called nuclear magnetons:

$$\mu_{яд} = eh / 2m_p,$$

where e is the absolute value of an electron's charge; h Planck's constant; m_p the mass of a proton. The nuclear magneton is by 1836.5 times less than the magnetic moment of an electron, so, the magnetic properties of atoms are caused by the magnetic properties of its electrons.

Several models of the structure of the atomic nucleus are known, which contradict each other in various relations. E.g., Bohr and Frenkel (1894–1952) offered a *liquid-drop model* of the atomic nucleus in 1936. According to their concept, the nucleus is similar to the drop of a charged dense liquid containing protons and neutrons strongly interacting among themselves. As well as in a usual liquid, the *liquid-drop nucleus*' surface is in a vibratory state, which may lead to disintegration (division) of the nucleus under certain conditions. In 1950, the American physicist Heppert-Mayer (1906–1972) and the German physicist Jensen simultaneously developed a shell model, in which the nucleons move independently from each other in an average field of the nuclear forces along certain shells (similarly to electrons) characterized by energy constants. Further, a number of scientists (O. Bohr, Mottelson, and Rainwater) presented a *generalized model* of the atomic nucleus, according to which the nucleus consists of a core (a steady internal part made of shells filled with nucleons) and so-called external nucleons moving in the field induced by this core. Finally, all his creates preconditions for the occurrence of strong vibrations leading to various deformations and even disintegration of the nucleus.

Techniques to measure the sizes of atomic nuclei are mainly base on observation of scattering of various particles (more often, alpha particles) by atomic nuclei. The distance between an alpha particle and the nucleus has been shown to be directly related with the scattering angle so (by a certain formula) that the distance between the centers of the particle and nucleus is equal to the sum of their radii at the instant of their contact. The radius of alpha particles can be easily determined by studying their scattering on helium nuclei. However, such a method applies to the nuclei of the lightest elements only with a small nucleus charge. In other cases, the energy of an alpha particle appears to be insufficient for overcoming electrostatic repulsion. Therefore, alpha particles, protons, and some other particles accelerated to high energies are used for measurement of the sizes of the nuclei of heavy elements. In particular, neutrons are more convenient for nuclei bombardment since, due to the absence of an electric charge, they can freely approach any nuclei and even get into them. The results of numerous experiments show that the sizes of atomic nuclei of all elements are 10^{-15}–10^{-14} m of the order of magnitude, which are less than the sizes of atoms by ten thousand times.

The following interesting in both scientific and practical relations phenomenon are associated with the structural specificity of the properties of atomic nuclei.

1) *Mass Defect.* From all (even inconsistent) theories of the atomic nucleus structure, it should follow that the nucleus' mass is equal to the sum of those of the constituent protons and neutrons. However, experimental data point to the true mass of a nucleus being always less than this sum. The difference between the theoretical and experimental values of the nucleus masses is called as the *mass defect.* The matter is that the process of formation of an atomic nucleus is interaction of the nucleons by their movement meetwards and their subsequent collision. Both particles (the proton and neutron) possess a high kinetic energy which can go to either overcoming of the nuclear attraction forces or its radiation (usually as a photon). Such energy release must occur at joining of each subsequent nucleon to the nucleus. Then, according to Einstein's formula $E = mc^2$, the energy reduction (as a result of its release at nucleus formation) must lead to reduction of its mass. This constitutes the physical sense of the mass defect phenomenon. In general, according to the substantive provisions of Einstein's theory of relativity, the mass change occurs at any processes conjugated with energy transmission. However, at usual energy values the mass change is often imperceptible. The energy values corresponding to nuclear interactions are rather high, and the mass changes are considerable in these cases.
2) *Radioactivity of chemical elements.* In 1896, the French physicist Becquerel discovered the phenomenon of radioactivity of chemical elements (with radiation by a uranium salt possessing high penetrability as an example). Further, it was established that the physical nature of this phenomenon consisted in spontaneous transformation of an unstable isotope of one element into an isotope of another one, which is accompanied by emission of electrons, protons, neutrons, or helium atomic nuclei (alpha particles). Of the elements contained in the earth crust, almost everything with a serial number above 83 possesses radioactivity. Natural radioactive isotopes disintegrate to release either alpha or beta particles. In 1934 the French physicists Iren and Frederic Jolio-Curie discovered the phenomenon of artificial radioactivity and carried out a nuclear reaction to form the radioactive isotope of phosphorus with its mass number 30. In 1940, the Soviet physicists Flerov and Petrzhak found spontaneous division of atomic nuclei. In general, division of atomic nuclei is a special process characteristic of the heaviest nuclei only, it usually occurs under the influence of neutrons and has the character of nuclear reaction. The process of division itself consists in disintegration of a heavy nucleus onto two parts with approximately equal masses and charges; and this process is often accompanied by emission of fast neutrons and gamma quanta.

It has been also established that at radioactive disintegration having a chain character, energy whose value exceeds the energy of chemical processes by hundred thousand and even millions times is released. In spite of the fact that the average lifetime of an atomic nucleus is a random value, for every radioactive substance there is its quite certain value, and the rate of decay obeys the exponential law:

$$N_t = N_0 e^{-\lambda t},$$

where λ is the radioactive disintegration constant; N_0 the initial number of unbroken atomic nuclei; N_t the number of unbroken nuclei at any instant t.

On the basis of numerous studies, a theory of chain nuclear reactions has been developed and the possibility of carrying out thermonuclear synthesis has been proven. Calculations show that at division of one nucleus the energy of an order of magnitude of 200 MeV (the kinetic energy of the nucleus' splinters and gamma radiation contribute 165 and 35 MeV, respectively) is released. In particular, at full division of one kg of uranium-235 the energy by few millions times higher than at burning of one kg of coal or oil is released. Besides, it has been established that, as at small distances between atomic nuclei the forces of attraction act, this may lead to synthesis of a heavier nucleus.

At high temperatures (several thousand degrees) such synthesis is called as thermonuclear, one of whose features is release of very high energy per mass unit of the reacting substances. E.g., participation of one gram of hydrogen isotopes in thermonuclear synthesis is equivalent to combustion of 10 tons of gasoline. Now, the man, in essence, owns only uncontrollable thermonuclear process (explosion of the H-bomb).

To implement controllable thermonuclear synthesis it is necessary, first, to heat up substances to even higher temperatures and, second, to create such conditions under which much more energy would be released at synthesis than has been spent for heating the substance.

3.9.7. Methodological Principles of Studying Nature in the Modern Physics

From the above considered concepts it is possible to draw a conclusion that the methodology of relativistic and quantum physics have their specificity in comparison with classical physics.

The methodological principles of modern physics can be formulated as follows.

1. The presence of objective existence of the material world (i.e. outside and irrespective of human consciousness), having, first of all, physical nature. This does not contradict the methodology of classical physics since this principle is most general for the development of all representations in physics.
2. The existence of three interrelated structural levels of the matter organization, namely, the microworld, macroworld, and megacosm, with the fundamental laws of nature (quantum physics, classical natural sciences, the general theory of relativity, respectively) acting in each of which.

3. The character of cause-n-effect relationships in the microworld essentially differs from mechanistic determinism; in microprocesses the causality principle is shown through the presence of many accidents (probabilities) and, consequently, statistical (instead of dynamic) regularities correspond to the microworld.
4. The description of the structure, properties, and functions of physical objects depends on the conditions of cognition. In relativistic physics it is due to the necessity to provide for the choice of the reference system (which does not contradict the constancy of the velocity of light in vacuum). In quantum physics it is shown in the interaction of the device used in the experiment with the microobject under study. This last statement is formulated in a generalized form as Bohr's complementarity principle.
5. Unlike classical physics whose methods allow determining many properties of a physical system simultaneously, in quantum physics the uncertainty principle acts.

The theories of modern physics contain not only elements necessary for the description of the properties of the studied objects and phenomenon mechanisms but also those describing the cognition process conditions.

REFERENCES

Einstein A., L. Infeld. *The Evolution of Physics*. Cambrodge University Press, 1938.

Gryadovoy D.I.. *Concepts of Modern Natural Sciences*. Moscow, 2003. [Russ].

Fermi E.. *Notes On Quantum Mechanics*. The University of Chicago Press, 1960.

Schrödinger E.. Does the Environment Determine Natural Sciences? / New Ways in Physics. Papers, speeches. Moscow, 1971. P. 22–45.

Wigner E.P.. Symmetries and Reflections: Scientific Essays. Indiana University Press, Bloomington and London, 1967.

Prigogine I.. From Being To Becoming. Freeman, 1980.

Landau, L.D. E.M. Lifshitz. *Course of Theoretical Physics*. Addison-Wesley, Butterworth-Heinemann, Pergamon Press, 1951–71.

Born M.. *Atomic Physics*. Blackie and Son, London, Glasgow, 1963.

Gliozzi. M. *Storia della Fisica*. Storia delle scenze, Torino, 1965.

Goldfein M.D., A.V. Ivanov, A.N. Malikov. *Concepts of Modern Natural Sciences*. Moscow, 2009. [Russ].

Feynman R.P.. *The Character of Physical Law*: The 1964 Messenger Lectures. MIT Press, 1967.

Feynman R.P.. *The Feynman Lectures on Physics*: The Definitive and Extended Edition. 3 volumes (2nd ed.). Addison Wesley, 1970.

Hawking S., R. Penrose. *The Nature of Space and Time*. Princeton University Press, Princeton, New Jersey, 1994.

Kuhn T.S.. *The Structure of Scientific Revolutions*. University of Chicago Press, 1962.

Kuznetsov V.I., G.M. Idlis, V.I. Gutina. *Natural Sciences*. Moscow, 1996. [Russ].
Ebeling Y.W., A. Engel, R. Feistel. *Physik Volutionsprozesse*. Berlin, Akademie-Verlag, 1990.
Vladimirov Yu.S.. *Fundamental Physics, Philosophy, and Religion*. Kostroma, 1996. [Russ].
Einstein. A. *The Collected Papers of Albert Einstein*, Vol. 1–10. Princeton University Press,1987-2006.

Chapter 4

THERMODYNAMICS AS A BASIS OF INANIMATE AND ANIMATE NATURE

Classical thermodynamics is the only physical theory of universal content which I am convinced will never be overthrown, within the framework of applicability of its basic concepts.

Albert Einstein

The concepts of classical and quantum physics allow one, either exactly or with a certain probability, to predict the state of macrobodies or microparticles. In particular, this concerns mechanical movement regularities which can be described by using spatial-temporal coordinates, the vales of mass, velocity, pulse, wave characteristics, the knowledge of the fundamental type of interaction. However, there exist some processes whose features can be explained by neither classical physics nor quantum representations. E.g. the existence of bodies to them in different aggregation states, the appearance of elastic forces at deformations of systems, possible transformation of some compounds into others, etc. As a rule, these and similar processes are accompanied by transition of systems from one state to another one with changes in thermal energy. Just such processes and most general thermal properties of macroscopic bodies are studied by the section of physics and chemistry called *thermodynamics*.

Thermodynamics studies a system, i.e. a body or group of bodies capable of changing under the influence of physical or chemical processes. Everything surrounding the system forms the external medium (environment). By a "body" they usually understand any substance having a certain volume and characterized by some physical properties. Thermodynamic systems can be homogeneous and heterogeneous. *Homogeneous* are such systems whose properties in macroscopic relation are identical at all points or change continuously upon transition from one point to another. *Heterogeneous systems* are made of elements different in their properties and separated with interfaces, on which there is a sharp change of properties. Every system is related with such a notion as the condition of a system, which represents a set of the quantitative values of all its thermodynamic properties. At change of at least one of these properties the system passes from one state to another one. Usually, thermodynamic processes proceed at constancy of certain parameters of the system. An *isothermal* process proceeds at constant temperature; an *isobaric* one proceeds at a constant external pressure; an *isochoric* one is characterized by the constancy of volume; an *isentropic* one features the constancy of the entropy of the system; an *adiabatic* one goes in the absence of thermal energy exchange between the system and its environment.

Energy is an important notion, which is distinguished according to the types of movement and interaction. In mechanics this is the energy of forward, rotary or oscillating motion of bodies; the energy of bodies' location in a force field (the potential energy in the gravitational field, in the field of electric forces, in the field of magnetic forces); the energy of elastic deformation (stretching, compression, and shift). In applied physics it is necessary to distinguish mechanical energy, the potential energy of gravity, the internal energy of heated bodies, acoustic energy, radiant energy, electric and magnetic energy. In technical and economic classification of energy types, the man-to-nature relation is usually reflected: the energy of solar radiation, hydraulic energy, wind power, the energy of fuel, mechanical energy, the internal energy of steam and compressed gases, chemical energy, electric energy, muscular energy of the man and animals. In thermodynamics, two main types of energy are distinguished, namely, external and internal ones. The notion of *external energy* covers the energy of movement of a body (system) as a whole and the energy of the body's location in a field of forces, when moving in which the thermodynamic state of the body does not change. The *internal energy* of a system is the sum of the energies of movement and interaction of all the elements constituting the system, i.e. the sum of the energy of forward and rotary motion of the molecules and oscillating motion of the atoms, the energy of molecular attraction and repulsion, intramolecular chemical energy, the intratomic energy of optical levels, intra-nuclear energy.

The bases of classical thermodynamics have been formulated in the works by Clausius, van't Hoff, Arrhenius, Gibbs, Helmholtz, le Chatelier, et al., which describes processes depending on changes of the properties of systems in an equilibrium state. In other words, the thermodynamic description of systems and phenomena is based on the idea of an equilibrium state. Thermodynamics answers the question: where a process is directed before equilibrium is reached. Therefore it does not deal, in an explicit form, with time as a physical parameter and considers no mechanisms of processes. If a system infinitely slowly passes from one state to another one through a continuous series of equilibrium states and if the maximum work is thus done, such a process is called *thermodynamically equilibrium process*. Between two next states of equilibrium, the values of any functions of state of the system differ by an infinitesimal value, and it is always possible to return the system to its initial state by infinitesimal changes of the functions of state. The system having made an equilibrium process can return to its initial state, having passed the same equilibrium states in the opposite direction, as in the direct one. Therefore, as a result of a *thermodynamically reversible process* the system and environment return to their original state, i.e. no changes remain in the system and environment. On the other hand, a process after which the system passes through a series of non-equilibrium states is called *thermodynamically non-equilibrium process*. If the system and environment cannot return to their original state, i.e. any changes remain in them, such a process is considered *thermodynamically irreversible*.

Equilibrium thermodynamics is, first of all, associated with molecular and kinetic theories about the structure and properties of macrosystems (e.g., the molecular and kinetic theory of gases). The main concept of such theories is that any body (gaseous, liquid or solid) consists of a large number of molecules being in chaotic movement; the intensity of this movement depends on the temperature of substance, and *vice versa*. That is why the structure, properties, and behavior of macrosystems are determined by the movement and interaction of the particles which the bodies consist of. There are quantitative relations describing molecular and kinetic features of the properties and behavior, in particular, gases, namely, the laws of Boyle– Marriott, Hay Lussac, Charles, Clapeyron–Mendeleyev, van der Waals, and some others.

At the same time, the scope of thermodynamics is much wider than the molecular and kinetic theory of gases. The matter is that, unlike many provisions of physics and chemistry, thermodynamics does not describe the course of processes in time (i.e. their dynamics), and the structure features of the studied system. In other words, thermodynamics only considers the initial and final states of a system which are characterized by special thermodynamic parameters. Usually, temperature, pressure, volume, and such *characteristic functions* as enthalpy, entropy, internal energy, Helmholtz' free energy, Gibbs' thermodynamic potential act as such parameters of state. Just these functions and their values characterize the thermodynamic state of a system and cause the probability of proceeding of this or that process accompanied by changes of thermal energy. The functions of state allow establishing the direction of spontaneous processes and determining the degree of their completeness in real thermodynamic systems. E.g., Gibbs' thermodynamic potential G can be used for exploring equilibrium (and quasi-equilibrium, as will be shown further) processes and closed (quasi-closed) systems, in which transformations proceed at constant temperature and pressure. Helmholtz' free energy F is applicable to studying similar processes and systems at constancy of temperature and volume.

Thermodynamics also features being a phenomenological science. This means that all physical quantities, functions and laws of thermodynamics are based on experience only, i.e. there is no strict theory concerning them. The sense of then empirical provisions of thermodynamics can be characterized as follows. 1) Unlike mechanical movement, all spontaneous thermal processes are irreversible (and, first of all, this applies to heat propagation in the environment). 2) According to various forms of movement and various types of energy there are also various forms of energy exchange. In classical thermodynamics, only two forms of energy transfer are resolved, namely, work and heat. Any kind of energy is an unequivocal function of the state of the system, i.e. the change of the energy does not depend on the way of the system's transition from one state to another one. At the same time, heat and work are unequal forms of energy transfer as work can be directly used for replenishment of the stock of any kind of energy, whilst heat – on replenishment of the stock of the system's internal energy only. 3) Any physicochemical system left to itself (i.e. in absence of external forces and fields) always aspires to pass to the state of thermodynamic equilibrium; this state is characterized by the uniformity of distribution of temperature, pressure, density, and the concentration of components. 4) The characteristic functions of the state of system are interrelated by certain ratios, for example:

$$\Delta F = \Delta U - T\Delta S; \qquad \Delta G = \Delta H - T\Delta S,$$

here ΔU, ΔH, and ΔS are changes of the internal energy, enthalpy, and entropy, respectively; T is absolute temperature, which points to their interconditionality and universality relative to the description of both initial and final state of the system.

The physical sense of the concepts of classical (equilibrium) thermodynamics is directly related with its three basic laws which are general laws of nature. *The first law of thermodynamics* in a qualitative sense coincides with one of the fundamental laws of nature, namely, with the law of energy conservation. In application to thermodynamics, its formulation is as follows: the quantity of heat given to a system (body) goes to the increase in its internal energy and to delivering work (A) by it (them):

$$\Delta Q = \Delta U + \Delta A.$$

The second law of thermodynamics has several formulations. E.g., any action associated with energy transformation (i.e. with energy transition from one form into another one) cannot occur without its loss as heat dissipated in the environment. More generally, this means that the processes of energy transformation can occur spontaneously only provided that energy passes from its concentrated (ordered) form to a diffused (disordered) one. One more definition of the second law of thermodynamics is directly related with Clausius' principle: any process at which there are no changes except heat transfer from a hotter body to a colder one is irreversible, i.e. heat cannot spontaneously flow from a colder location to a hotter location.. Such energy redistribution in the system is characterized by a quantity which has been named as *entropy*, which, as a function of state of the thermodynamic system (a function having a full differential) was first introduced by Clausius in 1865. Therefore, the more energy irreversible dissipates as heat, the higher entropy is. Whence it follows that any system whose properties change in time aspires to an equilibrium state at which the entropy of the system takes its maximum value. In this connection, the second law of thermodynamics is often called the law of increasing entropy, and entropy (as a physical quantity or as a physical notion) is considered as a measure of disorder of a physicochemical system. Most often, the system's transition from one state to another state is characterized by not the absolute value of entropy (S) but by its change (ΔS), which is equal to the ratio of the change of the quantity of heat (given to the system or taken away from it) to the absolute temperature of the system: $\Delta S = \Delta Q/T$, J/deg. It is the so-called thermodynamic entropy. Besides, entropy has a statistical sense since the number of those microstates by which a given macrostate is realized can serve as a quantitative characteristic of the thermal state of the system. In these cases, entropy is calculated by Boltzmann's formula (1844–1906):

$$S = k \cdot \ln W,$$

where W is the number of the microstates to realize this macrostate of the system; k is Boltzmann's constant ($1.38 \cdot 10^{-23}$ J/K).

Therefore, upon transition from one macrostate to another one, the statistical entropy also increases since such transition is always accompanied by a large number of microstates, and the equilibrium state (to which the system aspires) is characterized by the maximum number of microstates.

In connection with the notion of entropy, the notion of time gets a renewed sense in thermodynamics. As was already told, the direction of time is ignored in classical mechanics, and it is possible to determine the state of a mechanical system both in the past and in the future. In thermodynamics, time acts as an irreversible process of entropy increasing in the system. I.e. the higher the entropy, the longer time interval has been passed by the system in its development.

It is also necessary to mean that there are four classes of thermodynamic systems in nature:

a) isolated systems (upon transition of such systems from one state in another one no transfer of energy, substance, and information through the system's borders proceeds);
b) adiabatic systems (no heat exchange with the environment only);
c) closed systems (no substance transfer only);
d) open systems (which exchange substance, energy, and information with the environment).

Such a classification of thermodynamic systems, at first sight, leads to a seeming contradiction between the general principle of relativity in nature and the validity of the second law of thermodynamics for whole nature. The matter is that, ignoring exchange processes, the second law of thermodynamics is directly applicable to systems a) and b) only, for which entropy either is constant, or (which is much more often) increases: $\Delta S \geq 0$. This means that in such systems, the changes are characterized by increasing disorder (chaos) in them until the system will not reach thermodynamic equilibrium (or its entropy will not reach its maximum value). At the same time, some scientists think that, as the whole Universe as an isolated system obeys the second law of thermodynamics and the entropy of all its parts always increases (due to the infinite accumulation of diffused thermal energy in them), the state of the Universe aspires to chaos and, finally, the so-called thermal death must come. Such a gloomy (but wrong) conclusion is connected with two circumstances, namely: first, the notion of isolated system (as a nature element) is always an abstract notion; second, it was long ago established that entropy increases not in all systems, in particular, in open systems it may decrease. The latter fact, first of all, concerns biological systems, i.e. living organisms who are open non-equilibrium systems. Such systems are characterized by gradients of the concentration of chemicals, temperatures, pressure, and other physical and chemical quantities. In this regard, a seeming contradiction appears between the second law of thermodynamics and biological ordering.

The recovery from such a contradiction can be found in the use of the concepts of modern (i.e. non-equilibrium) thermodynamics which describe the behavior of open systems always exchanging energy, substance, and information with the environment. It should be noted that such exchange processes are characteristic not only of physical or biological systems but also of social-economic, cultural-historical and humanitarian systems since the processes occurring in them are, as a rule, irreversible. Schematically, the mechanism of interaction of an open system with the environment can be represented as follows. First, such a system at its functioning takes substances, energy, and information from the environment, and then gives it the used ones (naturally, in other types and forms). Second, all these processes are also accompanied by entropy production (i.e. energy losses and the formation of various wastes). Third (and this is most important), the formed entropy (characterizing a certain degree of disorder), without collecting in the system, is released into the environment, and the system takes from it new substances, energy, and information necessary for its optimum performance. All this allows drawing a conclusion that the open system, as a result of exchange processes, can be in a non-equilibrium state as well, for whose description new laws (in particular, non-equilibrium thermodynamics) are necessary.

Briefly, the creation history of the non-equilibrium thermodynamics concepts is represented as follows. In the first half of the 19th century, Fourier, Stocks, Navier, and Fick derived kinetic equations for the description of such non-equilibrium processes as heat conductivity, viscous liquid flow, and diffusion. Further, methods of the thermodynamics of non-equilibrium (irreversible) processes were used in analysis of the thermoelectric phenomena (W. Thomson) and electrochemical processes (Helmholtz). Onsager's works contributed much to the development of the thermodynamics of irreversible processes, which first introduced the notion of thermodynamic force. Just this quantity characterizes the degree of deviation of the system from its thermodynamic equilibrium and is the cause of irreversible changes in the system. The quantitative characteristics of the corresponding irreversible phenomena are called thermodynamic flows and, according to Onsager's representations, are the derivatives of the parameters of thermodynamic state with respect to time (*thermodynamic rates*). The concepts of the linear thermodynamics of non-equilibrium processes, offered by Onsager, describe the behavior of molecular systems at a phenomenological level and establish a relation between the kinetic factors determining the intensity of cross processes of transfer of heat, masses, momentum, and chemical reactions. Without pressing in further details, it is possible to draw a conclusion that the main objective of the thermodynamics of irreversible processes consists in formulating a set of equations describing the dependences of thermodynamic parameters on spatial-temporal coordinates on the basis of mechanics and the thermodynamics concepts (supplemented by the linearity and symmetry principles). The modern concepts developing non-equilibrium thermodynamics in its application to biological systems were developed by de Donder, de Groot, Denbigh, Mazur, Glansdorff, and Prigogine (it will be told about it in more detail below – in a chapter concerning the concepts of natural sciences in biology).

Another recovery from the specified contradiction is connected with the development of the concept of thermodynamics based on the possibility to apply classical (i.e. equilibrium) thermodynamics to real (open) systems. The matter is that many facts are now known to allow, in a certain degree, to neglect the irreversibility of some physical processes and chemical transformations (i.e. to study them in a certain approximation). In these cases it is merely necessary to establish such ranges of changes of the properties of a system, within which it is possible to consider them equilibrium ones. Usually, such states are called as quasi-equilibrium, when a process studied by changes of some parameter is steady and has such a time, in comparison with which the times of other processes in the system are either very long or negligible. Such a concept underlies *the law of temporary hierarchies* formulated by Gladyshev, which allows to resolve quasi-closed thermodynamic systems (subsystems) in open biosystems and to investigate their development (ontogenesis) and evolution (philogenesis) by studying changes in the value of Gibbs' specific (per unit of volume or mass) functions of formation of a given highest monohierarchical structure from the monohierarchical structures of a lowest level. Proceeding from these considerations, it is possible to make two important conclusions:

1) in every specific case, depending on the chosen time scale, (or process) this state of the system can be considered (with various degrees of the accuracy of approx.-imation) equilibrium or non-equilibrium, i.e., to apply various ways of description depending on the character of the task to be solved;

2) the second law of thermodynamics is applicable to real (open) systems as well, provided that the whole system (continuum) "open system–environment" is considered; for such a set, the total entropy will always increase, as for the first component $\Delta S_1 < 0$, for the second component $\Delta S_2 > 0$, and at $|\Delta S_1| \ll |\Delta S_2|$ we have:

$$\Delta S = (\Delta S1 + \Delta S2) > 0.$$

The third law of thermodynamics is associated with the notion of absolute zero. The physical sense of this law presented in Nernst's theorem consists in the basic impossibility to achieve absolute zero (–273.16°C) at which the forward thermal movement of the molecules must stop, and entropy will cease to depend on parameters of the physical state of the system (in particular, on the change of thermal energy).

REFERENCES

Gelfer Ya.M.. *History and Methodology of Thermodynamics and Statistical Physics*. Moscow, 1969. [Russ].

Gladyshev G.P.. *Thermodynamics and Macrokinetics of Natural and Hierarchical Processes*. Moscow, 1988. [Russ].

Goldfein, M.D., A.V. Ivanov, A.N. Malikov. *Concepts of Modern Natural Sciences*. Moscow, 2009. [Russ].

Kariakin N.V.. *Bases of Chemical Thermodynamics*. Moscow, 2003. [Russ].

Rebans K.K.. *Energy, Entropy, Environment*. Tallin, 1984. [Russ].

Vukalovich M.P., I.I. Novikov. *Thermodynamics*. Moscow, 1972. [Russ].

Chapter 5

DEVELOPMENT OF NATURAL SCIENCE CONCEPTS IN CHEMISTRY

The constitutional formula of a substance is not only a symbolic expression of the composition but reveals ... the intimate nature of that substance.

Jacobus Hendricus van't Hoff

5.1. MAIN NOTIONS AND PROBLEMS OF CHEMISTRY

Various types of matter may be chemical elements and their compounds, called substances, which have special importance in natural sciences. For astrophysics, of interest are such substances which stars and other heavenly bodies consist of and those existing in interstellar space. Biology explores the substances containing in the composition of live beings and actively participating in vital processes in organisms. In geology, main attention is paid to studying the substances which the planet Earth (minerals, ores, etc.) consists of. Physics is engaged in research of the physical properties of substances at both macrolevel and microlevel to establish fundamental laws of nature. From whence it follows that no clear boundary between chemistry and other natural sciences can be easily drawn. Nevertheless, it is possible to present the following definition. *Chemistry* is the science about the composition, structure and properties of substances, and about reactions which some substances are converted into others in.

In contrast to engineering with the notion "material" being commonly used, the "pure substance" is considered to be the primary notion in chemistry, while mixtures and solutions (which can be homogeneous or heterogeneous) are secondary ones. Any material system consisting of substances can be described by using the notions of phases and components which it is formed of. A phase is a homogeneous part of a system separated from its other parts with physical borders (e.g., ice floating in a closed vessel; the content of this vessel counts three phases, namely, the solid, liquid and gaseous ones). Components are those substances which all the phases of a system can be formed of. The ratio of the numbers of phases and components can vary in various chemical systems. Water is the only component in the above example. A molecular solution of any water-soluble substance in water is one phase containing two components. The qualities of various substances are characterized by their properties. Physical properties are such ones which can be observed and studied without converting this substance into another one. Chemical properties determine the ability of a substance to participate in chemical reactions; therefore, the notion *reactivity* is commonly used, which depends on the conditions of the process of transformation of initial substances into final products. Temperature is one of the most important conditions. This quantity can be defined proceeding from its physical sense. First, temperature is a measure of the intensity of thermal movement of the molecules which a physical system consists of. Second, temperature is the property to determine the direction of thermal energy transition from one object to another one (heating or cooling). The reactions accompanied by thermal energy release ($\Delta Q > 0$) are called exothermic; those going with thermal energy absorption ($\Delta Q < 0$) are endothermic; third, according to the second law of thermodynamics, absolute temperature is the derivative of the thermal energy of a system with respect to its entropy: $T = dQ / dS$. Absolute temperature is therefore always positive and expressed in Kelvin's degrees. Just about two hundred years ago, it was noticed that at cooling og a certain amount of gas its volume regularly decreased. From a theoretical extrapolation of this dependence it followed that at further cooling of this gas its volume must vanish at a temperature -273.16°C. From whence the conclusion was drawn that such a temperature should be minimal in nature, i.e. absolute zero. In this regard, the English physicist William Thomson (subsequently Lord Kelvin) (1824–1907) offered a new scale of temperatures originating from absolute zero. The absolute thermometric scale allowed expressing all the three laws of thermodynamics in the simplest and convenient form. Nevertheless, in scientific research and, especially, in common life the centigrade temperature scale is frequently used, which was introduced by the Swedish astronomer Anders Celsius in 1742 (the freezing point of water is 0°C, the boiling point is 100°C). Besides, two more temperature scales are known, René Antoine Ferchault de Réaumur's (°R) and Fahrenheit's (°F):

$$[°R] = 1.25 \cdot [°C]; \; [°F] = 5/9 \cdot [°C]; \; [K] = [°C] + 273.15.$$

To measure the masses of elementary particles, atoms, and molecules, the *atomic mass unit* is accepted (Dalton) which is equal to one twelfth of the rest mass of an unbound neutral atom of carbon-12 in its nuclear and electronic ground state, i.e.

$$^{12}C : 1\ Da = {}^{1}/_{12}\ m(^{12}C) = 1.9927 \cdot 10^{-26}\ kg / 12 = 1.660538921(73) \cdot 10^{-27}\ kg.$$

The *relative atomic mass* is the mass of an atom expressed in Da. The relative atomic of an element is the average relative atomic mass of the natural mixture of its isotopes providing for their contents in the terrestrial conditions. Just these values are given in Mendeleyev's Periodic System of chemical elements. The mole ("mol") is the amount of substance of a system which contains as many elementary entities as there are atoms in 12 g of pure carbon-12, namely, $6.02214129(27) \cdot 10^{23}$ mol^{-1} (Avogadro's number N_A). The *molar mass of substance* is the mass-to-quantity ratio of this substance, or the molar mass is the mass of one mole of the substance, i.e. the mass of its N_A structural particles.

Generalizing the majority of the problems facing chemistry, it is possible to formulate two main objectives, namely:

1) obtaining substances with preset chemical, physical and mechanical properties;
2) development of methods or ways to control chemical processes.

5.2. Formation of Chemistry As an Independent Science

One of the first stages of chemical knowledge development is associated with the concept of substance composition by Boyle (1627–1691). Its main idea is that the properties and qualities of material bodies depend on what chemical elements they consist of. The law of substance composition constancy (studies by Proust (1754–1826), Berthollet (1748–1822), and Dalton (1766–1844)) was a logic continuation of this concept. The sense of this law is in any individual chemical compound's possessing a certain and invariable composition, unlike mixtures and solutions.

According to this, any substance consists of such smallest particles (molecules) which determine its chemical properties. However, it is necessary to mean that the concept of substance composition constancy applies to low-molecular-weight compounds only, but has no exact scientific sense for high-molecular-weight ones due to some specifics of their formation mechanism (which will be stated in more detail in section 5.5).

The energetics of chemical transformations studied by thermodynamics is underlain by several laws of thermochemistry. Lavoisier (1743–1794) —Laplace's (1749–1827) law:

> The energy change accompanying any transformation is equal and opposite to energy change accompanying the reverse process (in particular, the thermal effect of formation of any compound is equal and opposite to the thermal effect of its decomposition). Hess' law: the enthalpy change for any chemical or physical process is independent of the pathway or number of steps required to complete the process provided that the final and initial reaction conditions are the same (the energy change accompanying any transformation is the same whether the process occurs in one step or many).

The notion "chemical equilibrium" is conceptual in chemistry; it is the state in which both reactants and products are present at concentrations which have no further tendency to change with time (reaction conditions, e.g. pressure, temperature, etc., do not change as well). Chemical equilibrium is reached when the forward and reverse reactions proceed at the same rate. Changes in the external parameters of a chemical system lead to violation (shift) of its equilibrium. Van 't Hoff's (1852–1911) law determines the direction in which equilibrium is shifted at a temperature change:

> if the temperature of a system at equilibrium changes, then at a temperature increase the equilibrium shifts towards the process going with thermal energy absorption, and at a temperature decrease — in the opposite direction.

Le Chatelier's (1850–1936) principle is more general as applies to much more types of equilibrium systems. In application to chemical processes it can be formulated as follows:

> if a chemical system at equilibrium experiences a change in concentration, temperature, volume, or partial pressure, then the equilibrium shifts to counteract the imposed change and a new equilibrium is established.

By the beginning of the 19th century, much experimental data were accumulated regarding the reactivity of various chemical compounds but which could not be explained from uniform positions. E.g., Gay-Lussac (1778–1850) established the existence of several (by chemical properties) organic substances of the same composition (which were called *isomers*). To explain various reactivity of substances, Berzelius (1779-1848) suggested considering a molecule as an ordered set of differently charged ions and radicals linked by electrostatic attraction (such a bond was further called *ionic*). Gerhardt (1816–1856) put forward a hypothesis of the molecule not being a simple set of atoms and radicals but a uniform and integral system whose main property was influence of these structural elements on each other. In the mid-19th century Kekule (1829–1896) formulated basic statements of his theory of valency to allow predicting the existence of new organic compounds and developing techniques of their synthesis. The creation of the Periodic Systems of chemical elements by Mendeleyev (1834–1907) in 1869 was a major achievement in the development of chemical knowledge. In it, all the elements known by that time were arranged in a certain order: at first, in the ascending order of their atomic masses, and further (after the discovery of the electron and atomic nucleus), in that of the atomic nucleus charge. Just such a concept allowed revealing the physical nature of the chemical element: the latter is a set of the atoms possessing an identical charge of the atomic nuclei.

However, despite the obvious progress in the chemistry development, none of the listed theories and concepts could provide a strict scientific justification of the main cause of the various reactivities of chemical compounds. The *theory of chemical structure* created by Butlerov (1823–1886) answered this major question. In this theory, main attention was paid to the distribution of *chemical affinity forces* among all the atoms in a molecule (called *chemical bonds*) rather than to just the spatial arrangement of atoms in this molecule. Just chemical bonds among atoms, possessing certain energies and lengths, have been shown to cause the various reactivity of substance.

Such a concept allows considering the chemical reaction to be redistribution of the chemical bonds in a reaction system (chemical bond breakage in the molecules of the initial reagents and of new chemical bond formation in the final products). At the same time, it has been established that the same chemical bond, depending on its surroundings (other atoms and bonds), is, as a rule, characterized by various strengths expressed by the value of bond energy. E.g., the C–H bond energy in the $H–CH_3$, $H–C(CH_3)_3$, $H–C(C_6H_5)_3$ compounds is 410, 335, and 70 kJ/mol, respectively. In a practical relation, the chemical bond concept allowed designing a large number of synthetics techniques for various substances and materials, and Butlerov's substance structure theory was the basis for creation of a new section of chemistry, namely, structural chemistry. The main concept of structural chemistry is that any substance consists of a set of molecules possessing a certain spatial structure.

From the viewpoint of modern natural sciences, the structure of molecules of low-molecular-weight compounds is a strict spatially and energetically ordered quantum system consisting of atomic nuclei and electrons.

At the same time, as well as the composition constancy concept, the structural chemistry of polymers has specific features. It is connected not only with features of the mechanism of their formation but also with the physicochemical and mechanical properties of high-molecular-weight compounds.

These are such properties as the statistical character of polymeric chain configurations, high elasticity, viscoelasticity, possible stereoregular structure, the ability to form copolymers, the possibility of their cross-linking together and getting block polymers, etc.

5.3. Some Basic Statements of Chemical Kinetics

For scientifically substantiated synthesis of various materials, it is not enough to know the composition and structure of the initial reagents and final products. It is very important to know the detailed mechanism of this or that reaction, which can be established on the basis of finding dependences of the kinetic parameters of the process in the whole and its separate stages on the reaction conditions (temperature, pressure, the nature of reagents, reactionary mixture composition, the acidity of the environment, etc.).

One of the basic statements of chemical kinetics is the law of mass action, i.e. the exponential dependence of the reaction rate W on the concentrations of the initial substances:

$$W = k \cdot C_1^{n1} \cdot C_2^{n2} \cdot \ldots ,$$

where C_1, C_2, etc. are the initial concentrations of the reagents (components); n_1, n_2, etc. the kinetic orders by components (can be either integer or fractional numbers); k the proportionality factor to show what rate the process would proceed with at the concentration of the reacting substances equal to unit, it is called the *rate constant* of a chemical reaction.

These kinetic orders can be found from the logarithmic dependence of the reaction rate on the concentrations of components (the straight line slope). As well as rate, the rate constant is one of the most important quantities in chemical kinetics, which greatest information on the reactivity of the reagents and the physicochemical specifics of the reaction is put in. This follows from Arrhenius' equation (1859–1927) for the rate constant:

$$k = A \times \exp(-E / RT),$$

where T is the absolute temperature of the reaction; R the gas constant; E the activation energy (the minimum energy to be additionally given to the reacting particles to involve them into the reaction); $\underline{A}$ the preexponential factor (a quantity containing some parameters determining the specific reagent participations in the reaction, e.g., the number of collisions between molecules, a steric factor, etc.). Depending on the reaction type (monomolecular, bimolecular or trimolecular), the dimension of the rate constant is s^{-1}, l/mol×s, l^2/mol^2×s, respectively. The value of activation energy can be found from the logarithmic dependence of the rate constant on inverse temperature ($\ln k$ vs. $1/T$).

As the initial concentrations of reacting substances do not depend on temperature, Arrhenius' equation holds true for the reaction rate.

$$W = W_0 e^{-E/RT}, \text{ where } W_0 = AC_1^{n1}C_2^{n2}.$$

Having differentiated this expression with respect to temperature, we get

$$d \ln W / dT = E / RT^2.$$

The last formula means that the relative increase in rate with temperature growth is characterized by the logarithmic derivative of rate with respect to temperature. Therefore, the higher activation energy, the quicker the reaction rate grows at temperature increase.

5.4. CONCEPT OF FREE RADICALS AND CHAIN CHEMICAL REACTIONS

Free radicals. Processes involving free radical have importance in physics, chemistry, biology, fuel industry, rocketry, and astronautics. The notion "radical" was first introduced into the chemical terminology by Lavoisier in 1789 to explain the structure of organic acids. After the discovery of free radicals and the development of techniques of their research, scientists acquired the possibility to study a new, chemically labile form of substance existence. Now, free radicals are known to be particles with uncoupled electronic spins (locating on boundary orbitals), possessing paramagnetism and high reactivity. Free radicals can exist in the gaseous, liquid, and solid states, be neutral or charged (radical ions), short-living (fractions of a second) or long-living (stable within several years) under usual (normal) temperature and pressure. Short-living radicals can exist much longer, but only at very low temperatures, in inert media or in vacuum. Burning and explosion, thermal cracking and polymerization, the processes of oxidation and stabilization of hydrocarbon fuel, lubricant oils, plastics, and rubbers, oxidizing destruction of foodstuff and medicines, various biochemical reactions belong to the reactions which free radicals participate in. Data on the structure and reactivity of radicals are obtained by means of modern physicochemical research techniques.

The application of spectroscopic methods to studying free radicals in a gas phase was effective. However, it was then established that the data obtained at the use of emission spectroscopy, often did not allow precisely establishing which radicals prevailed in the reaction zone. Absorption spectroscopy, mass spectrometry, and spin labels appeared to be more promising for research of real systems.

Along with various physical research methods for the structure and properties of free radicals, absolutely new kinetic ones based on the knowledge of the mechanism of chemical processes started to be applied in the early 1940s. These techniques were underlain by that circumstance that such kinetic characteristics of a chemical system as the reaction order by a component, activation energy, the stoichiometry of initiators and (or) inhibitors, etc., allowed estimating the rate constants of important stages of a complex chemical process. The introduction of the method of electronic paramagnetic resonance (EPR) into the chemistry of free radicals appeared most fruitful. The success of EPR in radical chemistry is caused by its direct studying the behavior of an uncoupled electron determining the paramagnetism of a free radical.

One of the most important features of EPR distinguishing it from other spectroscopic methods is that the total intensity of an absorption line (recorded at observance of certain conditions) allows direct determining the number of absorbing particles in the sample. This means that the spectroscopic absorption factors are identical for all radicals. EPR allows studying the magnetic characteristics of paramagnetic particles, which leads to a sharp increase in the sensitivity up to 10^{-10}–10^{-11} mol/l.

It is also necessary to note that when studying the fine and especially superfine structure of the EPR spectra of radicals and atoms it is possible to precisely establish the true structure of multiatomic radicals, specific features of their geometry, the density distribution of uncoupled electrons among various groups of a complex radical, and, in certain cases, even to obtain data on the interaction of radicals with their environment.

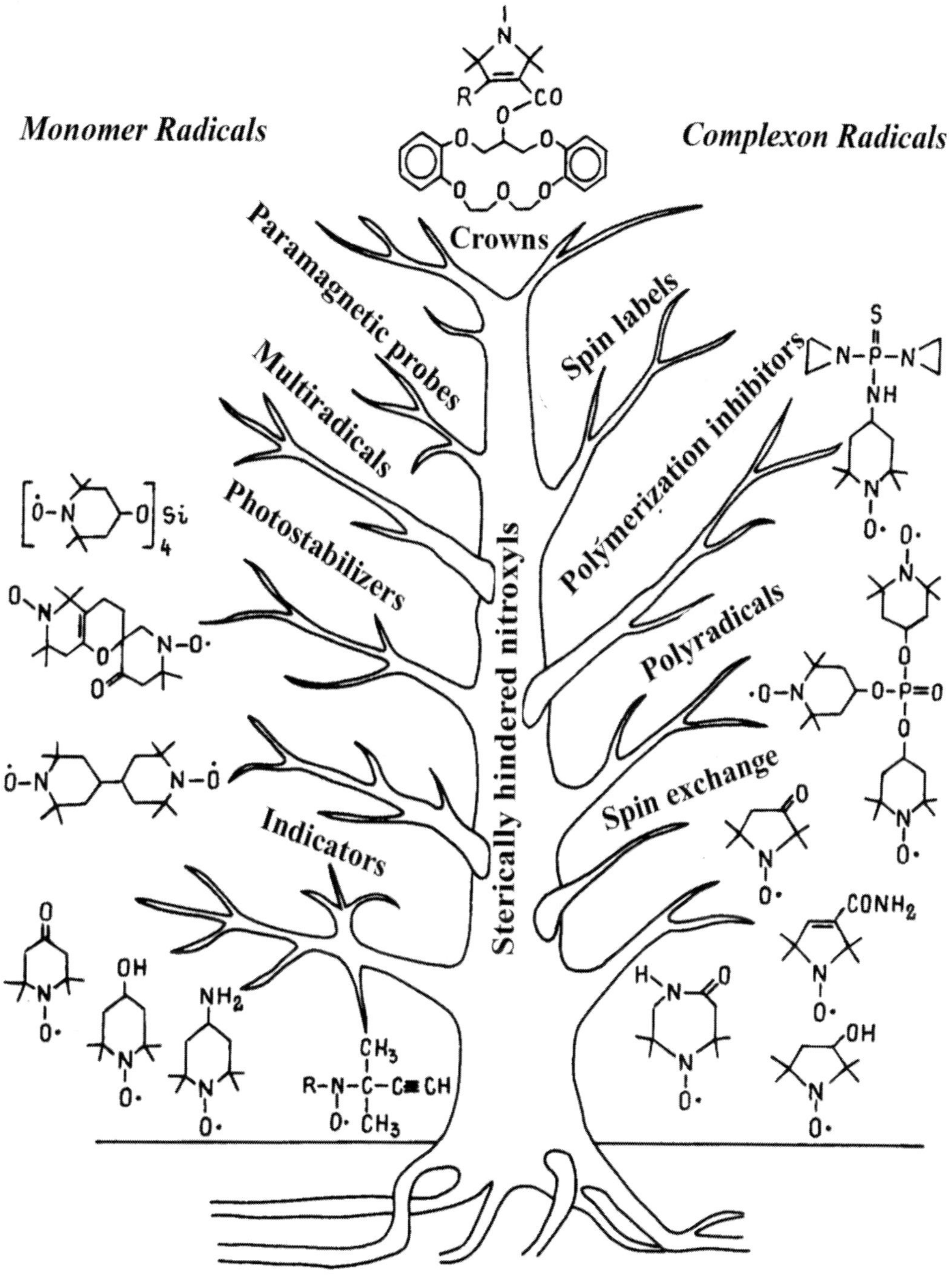

Figure 1. Genealogical tree of stable iminoxyl (nitroxyl) radicals.

Owing to organic chemistry success, chemically pure free radicals as substances stable in air have now become available at room temperature and under normal pressure. Almost all of them are brightly colored (red, orange, and dark-violet) solids or liquids.

Stable radicals are applied for intensification or retardation (suppression) of chemical processes, in biophysical and molecular biological research as *spin labels and probes*, in forensic medical diagnostics, when manufacturing films and other photographic materials, in instrument-making industry, to solve major environmental problems. Stable radical usage can be exemplified as follows.

A) Spontaneous reactions often proceed to form unnecessary polymers in the processes of synthesis, purification, and storage of monomers, which pollute the production equipment and lead to considerable loss of monomers. Aging of various polymeric materials is known to be often caused by their thermodestruction and photo-destruction. In all these cases, a large amount of wastes is formed, requiring their processing or utilization. Stable radicals, in particular, iminoxyl (I) and arnitroxyl (II) ones (Figure 1), have appeared to be very effective inhibitors of spontaneous polymerization, and to be stabilizers of polymers as well.
B) The properties of stable radicals allow making selection, assessment, and rejection of crystals and minerals. E.g., to evaluate the defectness degree of the surface of diamonds, a technique (EPR) has been offered based on the use of iminoxyl radicals as indicators to be detected directly in defect cavities.
C) Stable radicals have found application in the oil-extracting industry in control over oil deposit inundation. The essence of the method is that a solution of stable radicals is introduced into a forcing well, and then the reservoir liquid (oil with water) is regularly sampled. By the reduction of the radical concentration analyzed by EPR, the water increase in the oil field is traced.
D) Some iminoxyl radicals are fragments of complex compounds used as indicators of environmental pollution (in particular, these compounds change their color under the influence of radioactive radiation).
E) The essence of the method of spin labels for studying biological systems consists in that a stable radical covalently linked with an enzyme, polynucleic acid or protein plays the role of a peculiar tiny radio transmitter to continuously transfer information on the behavior of a biopolymer molecule. By analyzing the EPR spectra of spin-labeled biomolecules one can establish the protein structure, the nature of the active centers of enzymes, etc. A new diagnostic method of drug addiction is based on quantitative EPR determination of a spin-labeled drug (e.g., morphine or heroin) forced out from the cell by an unlabelled drug.
F) Some stable radicals have been established in special research to possess anti-cancer activity.

Chain reactions. As was shown above, free radicals actively participate in many chemical transformations and their free valency may either remain or disappear, depending on reaction type. The interactions of free radicals with the molecules of the initial reagents or solvent and their isomerization or disintegration appear to be the most probable processes. As a rule, a new free radical results, which, possessing high reactivity, enters the same reactions again to form a new free radical.

The sequence of such transformations will proceed until these free radicals, at some stage, meet with each other or are grasped by the reactor's wall (or by an inhibitor molecule – Inh).

Thermal decomposition (cracking) of ethane exemplifies it:

$$C_2H_6 \rightarrow 2\overset{\bullet}{C}H_3 \;\}\; \text{chain initiation}$$

$$\left.\begin{aligned}
&C_2H_6 + \overset{\bullet}{C}H_3 \rightarrow \overset{\bullet}{C_2}H_5 + CH_4;\\
&\overset{\bullet}{C_2}H_5 \rightarrow C_2H_4 + H;\\
&H + C_2H_6 \rightarrow H_2 + \overset{\bullet}{C_2}H_5;
\end{aligned}\right\}\ \text{chain propagation}$$

. . .

$$\left.\begin{aligned}
&C_2H_5 + C_2H_5 \rightarrow \text{inactive products};\\
&\overset{\bullet}{C_2}H_5 + \text{reactor's wall} \rightarrow \text{inactive products};\\
&\overset{\bullet}{C_2}H_5 + Inh \rightarrow \text{inactive products}.
\end{aligned}\right\}\ \text{chain termination}$$

Therefore, from the instant of C_2H_5 formation, a stage in the system begins, consisting of two alternating elementary reactions to convert C_2H_6 molecules into C_2H_4 and H_2 ones. *Such processes, in which the transformation of initial substances into final products proceeds by regular alternation of several elementary reactions with free radicals involved, are called as chain chemical reactions. Free radicals in these cases are active centers of these chain reactions.*

Theories of chain chemical reactions of various types were developed by Semyonov (1896–1986) and Sir Hinshelwood (1897–1967). Generally, a chain chemical reaction consists of three main stages, namely, initiation, propagation, and termination. *Chain initiation is the formation of free radicals from valently saturated molecules*, which may occur in various ways: monomolecular decomposition of initial substances, heterogeneous chain origin on the reactor's walls, photo- or radiation-chemical initiation, introduction of special substances, initiators (peroxides, azonitriles, etc.). *Chain propagation reactions are elementary reactions of a chain process in which the free valency of the active centers remains and which are accompanied by spending the initial reagents and the formation of the final products.* Usually, the stage of chain propagation consists of two or more elementary reactions and can be of four types: 1) the reaction of a free radical or atom with a molecule of one of the initial substances to form a new free radical; 2) the reaction of a free radical or atom with a molecule of one of the initial substances to form a molecule of the final product and a new free radical or atom; 3) the monomolecular transformation of a free radical in the chain into another one; 4) the monomolecular disintegration of a free radical to form a molecule of the reaction product and a new free radical or atom. *Chain termination is the stage of the process which leads to the death of the free valency of an active center.* Chain termination can have linear character when the rate of the elementary reactions of termination is directly proportional to the concentration of free radicals (death of radicals on the reactionary vessel's walls, interaction with valently saturated molecules, e.g., with inhibitor molecules). *Square-law* chain termination often occurs whose rate is proportional to the product of the concentrations of two interacting radicals.

Further, Nikolay Semyonov showed that reactions to increase the number of active centers (atoms or radicals of various chemical nature) proceeded in some chain processes along with elementary reactions of chain propagation. E.g., as a result of hydrogen oxidation, instead of one free valency (at an H atom), three ones are formed, namely, one at a hydroxyl and two at an oxygen atom:

$$H + O_2 \rightarrow \dot{O}H + \dot{O},$$

and the subsequent reaction of an oxygen atom with a hydrogen molecule leads to the formation of two univalent particles:

$$O + H_2 \rightarrow \dot{O}H + H.$$

Therefore, two new active centers emerge, i.e. one chain as though branches into three ones. So, such an elementary stage of a chain process leads to chain branching in which transformation of active intermediate products is accompanied by an increase in the number of atoms and free radicals of various chemical nature. Such reactions are *branched chain reactions*; in these cases the concentration of their active centers increases with time exponentially. A scientific lead has been developed in Semyonov's works associated with studying the processes of burning and explosion, in which two types of explosion or two types of spontaneous ignition (branched chain one and thermal) have been accurately differentiated. If chain branching prevails over chain termination, the reaction rate progressively grows in time and the reaction finishes with explosion almost without thermal emission (isothermal explosion). At the same time, along with chain ignition, thermal ignition is also possible, when the cause of an explosion consists in that the heat released in the reaction has not enough time to be completely removed into the environment and the sharply increasing temperature considerably increases the rate of the chemical process. Further, explosion regularities for autocatalytic-type reactions were established, thermal explosion criteria were established, i.e. the conditions of its emergence and proceeding in the form of various quantitative ratios between the duration of a chemical reaction and the thermal relaxation time, thermal explosion for the case of convective heat transfer was considered, etc.

Besides, the emergence of new chains is also probable from the final products of a chain reaction if they can form free radicals easier than the initial substances. For example:

$$ROOH \rightarrow \dot{R}O + \dot{O}H,$$

which leads to additional chain initiation. The reactions including such initiation are *degenerated branched* chain reactions. Their basic difference from the branched chain reactions consists in that new chains arise during a time considerably exceeding the period of chain existence whereas for usual branched reactions the formation of new chains occurs in the course of propagation of the chain itself. In particular, rather slow self-acceleration observed in certain cases at explosion is caused by the existence of degenerated branching.

At last, a theory to describe the role of the reaction vessel's walls in the processes of generation of active centers and chain termination (e.g., inverse proportionality of the termination reaction rate to the squared diameter of the reactor) has been developed.

On the basis of the foregoing it is possible to draw the following conclusions:

- the theory of chain chemical reactions is based on modern physical concepts of the structure of substance and energy transfer among reacting particles;
- the rate of all known chain chemical reactions depends on time (even at constancy of the concentrations of initial substances, which corresponds to the initial stages of a process);
- those chain reactions whose rate increases exponentially under certain conditions lead to explosion; this means that chain chemical reactions may not obey Arrhenius' law (e.g., at approaching the explosion temperature) in certain cases;
- the existence of critical phenomena caused by changes of various physicochemical factors (temperature, pressure, the composition of the reaction mixture, the size of the reaction vessel, the state of its walls, etc.) is characteristic of chain chemical reactions; in particular, such critical phenomena relative to pressure are the "lower" and "upper" limits of ignition, a sharp change in the reaction rate at a minor change of the inhibitor concentration, etc.;
- the rates of some chain chemical reactions and ignition areas may depend very strongly on the existence of very insignificant concentration (traces) of impurities in the reaction system.

5.5. High-Molecular-Weight Compounds

5.5.1. Some Features of the Thermodynamics and Kinetics of Polymer Formation Processes

The formation of a high-molecular-weight compound, first of all, is determined by the existence of unsaturated bonds, heterocycles, and multifunctional groups in monomers. Polymerization is a typical representative of chain chemical reactions, each of which consists of several main stages, namely, initiation, propagation, and termination. The type of chain polymerization is determined by the nature of active centers (intermediate particles) participating in each stage of the process. Ions (cations or anions) or free radicals can be such centers. Ionic and radical polymerization proceeds in the former and latter case, respectively. Breakage of one π-bond and the formation of two σ-bonds occur at radical polymerization. From whence, the thermal effect of the reaction can be found from the difference of the dissociation energies of the corresponding bonds: $\Delta H = 2E_\sigma - E_\pi$; this means that polymerization is an exothermic reaction since $E_\sigma > E_\pi$. In particular, ΔH is ca. 80 kJ/mol for many monomers of the vinyl row; its value depends on the nature and location of chemical bonds, and on the substituents in the main chain and their conjugation (interference).

Another necessary condition of polymerization is the thermodynamic possibility of monomer transformation into polymer. From the concepts of thermodynamics it is known that any chemical reaction proceeds with reduction of Helmholtz' free energy F (in isochoric-isothermal conditions) or with reduction of Gibbs' thermodynamic potential G (in isobaric-isothermal conditions of the system):

$$\Delta F = (\Delta U - T\Delta S) < 0;$$

$$\Delta G = (\Delta H - T\Delta S) < 0.$$

Similar changes are observed at polymerization as well, which is thermodynamically possible when $\Delta G < 0$. As the formation of a polymer is an exothermic process, $\Delta H < 0$. On the other hand, with molecular chain growth, the freedom of various types of movement of monomeric units decreases, which leads to an increase in the orderliness of the system and to entropy reduction. This means that $\Delta S < 0$, $-T\Delta S > 0$, and $|\Delta H| > |-T\Delta S|$. At polymerization, entropy reduction is generally due to retardation of forward (rather than rotary or oscillatory) movement. Therefore, the lower the temperature, the higher the probability of $\Delta G < 0$ is and polymerization is thermodynamically possible. At increasing temperature, the value of the entropy term increases and at some $T_\infty = \Delta H / \Delta S$ (different for each monomer) ΔG may vanish. At further increasing temperature, $\Delta G > 0$. Hence, when $T < T_\infty$ there are no thermodynamic obstacles for polymerization, whereas when $T > T_\infty$ monomer conversion into polymer is impossible. At the same time, it is necessary to mean that such a condition is necessary but insufficient since at calculation of Gibbs' thermodynamic potential one should consider not only temperature but also such physicochemical factors and processes as dissolution, crystallization, complex formation, intermolecular interactions of the polymer with the monomer and solvent, the nature and concentration of the reaction mixture components.

In the simplest form, the scheme of radical chain polymerization beginning under the influence of a substance specially introduced into the reaction environment (an initiator) can be presented as follows.

Chain *initiation* consists of two elementary reactions, namely, decomposition of the initiator I (usually on two radicals called primary ones) and addition of these primary radicals R_0 to a monomer molecule M to form a new radical R_1. Each of these elementary stages proceeds with a certain rate, and the reactivity of the particles involved is characterized by the rate constants of the reactions of initiator disintegration k_{dis} and that of polymerization initiation k_i. Initiation is the hardest stage (characterized by the highest activation energy) to limit polymerization in the whole:

$$\left.\begin{array}{l} I \xrightarrow{k_{dis}} 2R_0 \\ R_0 + M \xrightarrow{k_i} R_1 \end{array}\right\} \text{chain initiation;}$$

The second stage, chain *propagation* , is a sequence of elementary reactions of addition of active radicals R_1, R_2, R_3... R_{n-1} to monomer molecules to form active radicals as well. So, the molecular chain grows along with the kinetic chain propagation since every subsequent radical is longer than the previous one by one unit (a monomer molecule). Each of the elementary reactions of chain propagation is characterized by its own rate constant k_{p_1}, k_{p_2}, k_{p_3}, etc.

$$\left.\begin{array}{l} R_1 + M \xrightarrow{k_{p_1}} R_2\,; \\ R_2 + M \xrightarrow{k_{p_2}} R_3 \\ \text{..................................} \\ R_{n-1} + M \xrightarrow{k_{p_{n-1}}} R_n \end{array}\right\}\ \text{chain propagation;}$$

The third main stage (chain termination) usually consists of either one or two elementary reactions of interaction of polymeric radicals with each other to form one (P) or two ($P_1 + P_2$) polymer molecules. Naturally, the termination reactions proceed with considerably high rates (in comparison with chain initiation and propagation) and are generally characterized by the rate constant of the chain termination reaction k_t:

$$R_m + R_n \xrightarrow{k_0} P \text{ or } P_1 + P_2 \text{ – chain termination}$$

By using of chemical kinetic methods, it is possible to calculate the above radical chain scheme and to derive a general expression for the rate of the whole process W. Without going into calculation details, it is possible to make two conclusions:

- first, the absolute rate of radical chain polymerization is determined by the reaction rate of chain propagation

  $W = k_p \cdot [R] \cdot [M]$;

- second, the stationary concentration of polymeric radicals is

  $[R] = (W_i / k_t)^{0.5}$,

  where W_i is the initiation rate. Therefore, we finally get:

$$W = \frac{k_p (k_i)^{0.5}}{(k_t)^{0.5}} [I]^{0.5} [M]$$

Then, representing each rate constant through Arrhenius's equation, it is possible to get an expression for the effective activation energy of the process of polymer formation through the activation energy of the elementary stages, E_i (initiation), E_p (propagation), and E_t (termination):

$$E_{ef} = 0.5E_i + E_p - 0.5E_t.$$

As for the majority of monomers $E_i \approx 120$ kJ/mol, $E_p \approx 20$–30 kJ/mol, $E_t \approx 4$–8 kJ/mol, then $E_{ef} > 0$. And as E_{ef} is the temperature factor of the polymerization rate (or rate constant), from the inequality $E_{ef} > 0$ it follows that the polymerization rate always increases with temperature.

The polymerization degree P or the molecular chain length is an important quantity to characterize the physicochemical and mechanical properties of macromolecules. It can be calculated differently: or through the polymer-to-monomer molecular mass ratio $P = M / m$ or through the chain propagation-to-termination rate ratio:

$$\overline{P} = k_p[R][M] / k_t[R]^2.$$

Just from this expression it follows that the polymerization degree (or the molecular chain length, and, hence, the molecular mass of polymer) is always an average statistical quantity. The matter is that the propagation of each kinetic (and the growth of the corresponding molecular) chain occurs up to various lengths; so, chain termination occurs in a random way. Therefore, the same polymer is a set of molecules of various lengths (sizes), i.e. the molecular mass of polymer has polydisperse character. In such a case it is only possible to speak of the average molecular mass of polymer and, hence, the concept of composition constancy is meaningless for high-molecular-weight compounds.

Similarly to how the expression for the effective activation energy of polymerization has been derived, it is possible to get one for the effective activation energy of polymer formation with a molecular chain length P:

$$E_P = E_p - 0.5E_t - 0.5E_i.$$

At usual values of the activation energy of the elementary stages of polymerization $E_P < 0$; therefore, the molecular mass of polymer always decreases with polymerization temperature increasing at thermal and material initiation.

5.5.2. Features of the Structure and Properties of Macromolecules

The solid and liquid aggregation states are typical for polymers; they are characterized by oscillating and rotary motion of particles. Polymers do not exist in a gaseous state because it is necessary to overcome the strong intermolecular interactions among chain macromolecules to move apart macromolecules at big distances; this requires such energy expenses which are comparable with the energy of chemical bonds in a polymeric chain, which would lead to polymer destruction. Generally, phase state is determined by the order in particle arrangement. The thermodynamically stable *amorphous* state is characterized by near order at 1.0–1.5 nm distances and isotropic properties. Several models of the amorphous state of polymers are now known, namely, batch, statistically combined macromolecules (domains) and their sets (fibrils) and some others. The *crystalline* phase state is characterized by distant three-dimensional order in the arrangement of both chains and units.

By the structure type of chain molecules, high-molecular-weight compounds are classified on linear, branched, and cross-linked (or network) ones. Linear polymers represent a continuous sequence of identical atomic groups (M) extended in one direction, i.e. schematically as –M–M–M–M–M–. Polyethylene, polypropylene, polytetrafluoroethylene, polycaproamide (Capron™) are typical representatives of linear polymers. If one of the functional groups (radicals) is replaced by a lateral chain, a branched polymer results. Usually, such branched macromolecules are formed when the monomer contains three or more functional groups; they can be irregular structures, or comb-like or star-like. Network polymers have a three-dimensional spatial structure formed by chemical bonds. Natural rubber exemplifies such macromolecules; it represents (as a result of the curing process with the participation of sulfur atoms) a highly elastic network. The structural irregularity is characteristic of the majority of macromolecules. However, under certain conditions it is possible to obtain stereoregular polymers where all the carbon atoms possess an identical spatial configuration: i) isotactic polymers (all substituents are either above, *l*-form, or below, *d*-form of the main chain) or ii) syndiotactic polymers (macromolecules are built with an opposite spatial configuration, the *l*- and *d*-forms to alternate).

One more important property of macromolecules is their *flexibility*; the mechanism of it can vary. The main cause of polymer flexibility consists in the possibility of free rotation round unary C—C bonds. Owing to the symmetry of electronic density distribution, rotation around a σ-bond is possible with no change in the intensity of overlapping of the electronic clouds, i.e. without breaking of the chemical bond. No rotation without chemical bond breakage is possible round a double C=C bond. It is necessary to note that the property of flexibility is absent in low-molecular-weight compounds, despite the possibility of rotation of atomic groups. The matter is that for a long macromolecule, several turns are enough to reverse the direction of the chain and to realize configurations with various degrees of coiling. So, the configuration of a polymeric molecule is such a relative spatial arrangement of its units which can be changed as a result of reversible internal rotation without breakage of the chemical bonds. According to the second law of thermodynamics, owing to thermal movement, a rather long and flexible macromolecule is usually coiled into the most probable state with the maximum entropy, i.e. into a *statistical coil*. On the other hand, the most improbable state (with the minimum entropy) is the completely extended (rod-like) macromolecule, an ideally rigid chain in which no rotation is possible. Along with these structures, other types of macromolecular models are also possible, having absolutely other nature of chain flexibility. This can be exemplified by a DNA macromolecule representing a double helix in which a turn of one of the chains is limited by the influence of the other chain. Some more notions are associated with the property of flexibility of polymeric molecules, namely: *segments* (by means of whose moving their thermal movement occurs), the *high elasticity* and *fluidity* of macromolecules. The *segment* is a statistically independent piece of the chain whose location in space does not depend on other units. Experimentally, the length of a segment is estimated by studying of the thermodynamic properties of polymeric solutions. Except segment length, the *root-mean-square end-to-end distance* of a macromolecule is also a measure of the flexibility of a polymeric chain.

Owing to the chain structure and flexibility of macromolecules there are various physical states of amorphous polymers: *glassy*, *highly elastic*, and *plastic*. At low temperatures, oscillating motion of lateral atomic groups (radicals) only occurs and the polymer is in a glassy state; with temperature increase the stock of internal energy grows and rotational oscillations of the segments passing to their rotation become possible. The polymer passes to a highly elastic state characterized by segment mobility. Generally, *high elasticity* is the ability of bodies to high reversible deformations under the influence of rather small loads. From the molecular viewpoint, the essence of highly elastic deformation of polymers consists in straightening of the coiled flexible chains under the influence of the applied load and their return to the initial coiled state after loading removal. At further temperature increasing, viscous fluidity is observed, caused by the mobility of macromolecules as wholes. *Fluidity* is the ability to irreversible directed moving of macromolecules relative each other without violation of the body's integrity, and the systems possessing both fluidity and elasticity are called viscoelastic systems. Therefore, the upper temperature border of the highly elastic state is the temperature of fluidity T_f, and the lower temperature border is the temperature of vitrification T_v. Transition from one physical state into another one proceeds gradually, within a temperature range of 10–20°. The rheological properties determining the dependences between tension, deformation and its rate are important to optimize processing of polymeric materials. The specificity of flow in polymers consists in that long flexible macromolecules move owing to consecutive moving of their segments. If for low-molecular-weight liquids the activation energy of viscous flow (E_{visc}) grows with molecular mass, for macromolecules, beginning from the molecular mass of a segment M_s, E_{visc} remains constant and does not depend on further increase in the molecular chain length. At the same time, the value of E_{visc} depends on the nature and structure of macromolecules. One and the same polymer can pass to a glassy state at fast cooling and crystallize at rather slow cooling. From the viewpoint of physical kinetics, segmental mobility is "frozen" in the glassy state. At temperatures below T_v the energy of thermal movement appears insufficient to overcome the intra- and intermolecular interactions. A spatial network formed between next chains by cross-links of non-covalent nature (interaction of polar groups, H-bonds, etc.) appears in a polymer. Besides, for polymeric glasses, friable packing of macromolecules is characteristic, whose consequence is lack of *fragility* (polymer destruction under the influence of loading and the development of elastic deformations).

The ability to crystallization depends on the structure, flexibility, chain regularity, and packing density. The optimum crystallization temperature has the average value between the vitrification and melting temperatures ($T_v < T_{cr} < T_m$). The longer the crystallization time (a low cooling rate), the more regular equilibrium crystal structure is formed. The crystallization mechanism of polymers is almost the same as for low-molecular-weight substances. It consists in the formation of crystallization *centers* in an amorphous liquid phase (the first stage) and their growth (the second stage). They can arise either instantly at the initial instant of crystallization (then the number of centers is time-independent) or their number linearly depends on time.

The growth of centers can be one-dimensional, two-dimensional or three-dimensional to form rods, disks or spheres, respectively. Studies of the mechanical properties of crystal polymers have shown that crystallization can occur at rather high tensions as a result of structure reorganization at stretching (the formation of a "neck", a site with a reduced section).

5.6. A Short Characteristic of Laser Chemistry

Laser chemistry is a section of modern chemistry. It started to develop in the 1960s after the invention of lasers, devices to transform various types of energy (electric, light, chemical, thermal, etc.) into the energy of coherent electromagnetic radiation (oscillations described by harmonics, i.e., sinusoidal ones). The name "LASER" is an abbreviation of "Light Amplification of Stimulated Emission of Radiation". The operation principle of the laser consists, first, in transfer of the quantum system (a source of laser radiation) into an excited state (as a result of intense energy absorption by the atom of substance) by the influence of a power source on it (laser pumping) and, second, electromagnetic radiation emission by this excited quantum system. Depending on laser type, various substances are used as a source of generated radiation, namely, solid (glasses, rubies, and garnets) which compounds containing easily excited atoms are added in, and solutions of dyes or atoms of noble metals. By selecting the frequency of laser radiation, it is possible to change the depth of its penetration into the chemical reaction zone. The high uniformity (monochromaticity) of radiation has major importance to allow selective activating certain molecules in the reaction mixture. In other words, the mechanism of laser chemistry is in selective activation of certain chemical bonds by laser radiation which, finally, should lead to the formation of substances with preset properties. Such a mechanism is caused by the following. Any chemical bond in a molecule can be stretched or bent, i.e. be in a vibratory state characterized by a certain frequency. This frequency depends on the nature of atoms and on the location of the bond within a molecule (on its environment). Therefore, to break this bond it is possible to use a laser tuned to the characteristic frequency of this bond, i.e. to the frequency determining the specificity of the chemical bond. Then, a laser beam (like a guitarist's fingers touching a guitar string) starts to shake a certain chemical bond, which leads to its weakening and, finally, breakage. Thus, it is possible to draw a conclusion that laser chemistry is an effective method of initiation (or activation) of chemical processes.

References

Buchachenko A.L., A.M. Vasserman. *Stable Radicals*. Moscow, 1973. [Russ].

Emanuel N.M., D.G. Knorre. *Course of Chemical Kinetics*. Moscow, 1974. [Russ].

Goldfein M.D., A.V. Ivanov, A.N. Malikov. *Concepts of Modern Natural Sciences*. Moscow, 2009. [Russ].

Goldfein M.D., L.I. Karnaukhova. *Principles of the Physics of Synthetic and Natural Macromolecules*. Saratov, 1998. [Russ].

Goldfein M.D., N.V. Kozhevnikov, A.V. Trubnikov. *Kinetics and Mechanism of Regulation of Polymer Formation Processes*. Saratov, 1989. [Russ].

Goldfein M.D.. *Kinetics and Mechanism of Radical Polymerization of Vinyl Monomers*. Saratov, 1986. [Russ].

Gryadovoy D.I.. *Concepts of Modern Natural Sciences*. Moscow, 2003. [Russ].

Klenin V.J., I.V. Fedusenko. High-Molecular-Weight Compounds: A Textbook. Saratov, 2008. [Russ].

Nonhebel D.C., J.C. Walton. *Free-Radical Chemistry: Structure and Mechanism*. Cambridge University Press, 1974.

Pauling L.. *General Chemistry*. California Institute of Technology, Pasadena, 1944.

Reactivity, Mechanism and Structure in Polymer Chemistry / Eds. A.D. Jenkins, A. Ledwith. Wiley-Interscience, London, New York, 1974.

Rozantsev E.G., M.D. Goldfein, V.F. Pulin. *Organic paramagnets*. Saratov, 2000. [Russ].

Semyonov N.N.. *Chain Reactions*. Moscow, 1986. [Russ].

Volodarsky L.B. et al. *Imidazoline Nitroxyl Radicals*. Novosibirsk, 1988. [Russ].

Chapter 6

CONCEPTS OF NATURAL SCIENCE IN BIOLOGY

The laws of life are part or a consequence of some general law governing the evolution of matter in the whole.

Charles Darwin

6.1. FEATURES OF LIVE BEINGS. MAIN OBJECTIVES OF MODERN BIOLOGY

For many years, biology seemed to be the science of live substance (including live beings) to study its emergence, development, and death. From ca. the mid-20th century, as a result of certain achievements of the evolutionary, molecular and genetic biology, it became clear that the problems of live beings could be successfully solved at an essentially new qualitative level by applying an interdisciplinary approach. By now, biology has turned into a set of sciences of animate nature, including not only the questions of reproduction and development of organisms but, mainly, the problems of their variability, hereditary information transfer, metabolism, and adaptation to changing conditions of the environment.

From the viewpoint of modern natural sciences, *life* is the highest form of matter movement, realized in open systems of several organizational levels, which consist of both low-molecular-weight and high-molecular-weight compounds. Such systems constantly exchange energy, substance, and information with their environment and are characterized by self-reproduction and self-control processes.

The common properties of live beings to distinguish them from inanimate nature's bodies can be formulated as follows.

1. Organisms possess a complex but considerably more ordered structure, which is caused by the continuous inflow of substance and energy.
2. The existence of a genetic code is the basis of their ability to self-reproduction.
3. They actively react to the influence of both abiotic and biotic environmental factors.
4. They possess a high degree of adaptation to environmental changes.
5. Structural asymmetry is characteristic of them, which is exhibited not only at the macrolevel but also at the molecular and supramolecular ones (molecular chirality).
6. Organisms possess the ability to store and pass information to each other as well as inside themselves.

7. Biological systems use food as a source of energy and substances necessary to maintain all their activity processes. Main groups of the organic world of the highest taxonomical rank differ by their way of getting and processing food. Almost all plants are capable of photosynthesis, one of the forms of *autotrophic* nutrition. Animals and mushrooms eat differently, by using other organisms' organic substance.
8. Any biological system consists of separate parts closely interacting among themselves to form a structural-functional unity.

Besides, along with the general visible analogy, a basic distinction exists between the growth of live and lifeless systems. Inanimate nature's objects (crystals, stalagmites, etc.) grow by attaching new substance to their external surface. Live systems grow "from within" at the expense of nutrients gotten by the system in the course of autotrophic or heterotrophic nutrition. New live protoplasm is formed as a result of assimilation of these substances.

The main objectives of modern biology are:

1) studying regularities of the structure, properties, development, and functions of a cell, tissue, organ, and their systems, and separate live beings (including the man);
2) research of features of the organization of populations, species, biocenoses, and ecosystems;
3) establishment of the mechanisms of metabolism and hereditary information transfer;
4) research of regularities of the evolution processes of live organisms;
5) development of modern physical and chemical research techniques in the evolutionary, molecular and genetic biology, including the application of mathematical and physical modeling and new information technologies.

6.2. Principles of the Live Matter Organization

The simplest scheme counts the following levels of the live matter organization.

1. *Atomic and molecular level.* Seven chemical elements (organogenes), namely, carbon, hydrogen, oxygen, nitrogen, calcium, phosphorus, and sulfur constitute the basis of live beings; their mass fraction in biological systems is above 97%. Among these, a special place is occupied by carbon whose fraction in the crust is 0.1 wt. % only. The ability of carbon atoms to form strong covalent bonds with each other is one of the most important properties of them. Sixteen more elements are important for functioning of live beings, namely, sodium, potassium, manganese, magnesium, iron, aluminum, copper, zinc, cobalt, silicon, nickel, vanadium, molybdenum, selenium, iodine, and chlorine, their mass fraction being ca. 1.5%. At the same time, any live system consists of low-molecular-weight inorganic and organic substances (water, mineral salts, amino acids, simple and complex fats, enzymes, carbohydrates), and of biological macromolecules (polypeptides, polysaccharides, polynucleotides). Water molecules account for above 90 wt. % of live cells; their major importance for the development of organisms is caused by their ability to be solvent

in various biochemical reactions. Main vital functions, including metabolism and the transfer of hereditary signs and features, begin and proceed just at the atomic and molecular level.

2. *Cellular level.* The biological cell as an elementary particle of live matter is the simplest form of the existence of live beings. Just because of this, main results of biochemical research were obtained by studying cells and unicells. The biological cell is an ordered structural-functional complex capable of independent existing, breeding, and actively reacting to environment irritants. Biological cells absorb and transform nutrients, accumulate and use energy, store and transfer hereditary information. All endocellular structures and their functions are integrally interconnected, and biochemical reactions proceed in specially organized molecular structures. If a cell loses its ability to perform at least one vital biochemical function, it perishes at once. This elementary particle of live matter possesses all functional properties inherent in most complex multicellular organisms. These are such properties as reproduction, growth, metabolism, development, heredity, diseases, aging, and death inevitability. For all these properties, a so-called feedback is characteristic, manifesting itself in two interconnected processes. First, live cells consume substances and energy from the environment, and, second, they return certain quantities of processed substances and energy into it as vital activity wastes. This, in turn, means that open systems support and even often increase their orderliness, but, at the same time, bring a certain disorder into the environment. Just in this way, biological systems actively participate in the evolution processes on the Earth being a powerful factor of the formation of its separate components (landscapes, minerals, and ores). By internal and external signs and features, cells fall within two classes: 1) *prokaryotic* ones, which have no nuclei and form bacteria and blue-green algae; and 2) *eukaryotic* ones, which have been formed as a result of division of a unique impregnated cell (zygote) and which the highest plants and animals are built from. The new cells arisen in such a way were transformed to the tissues and organs of an embryo, which grew and developed as a result of the continuing reproduction and cell fission up to the formation of an adult organism. And the most important in these complex processes is the absolutely strict coordination of all their stages according to a certain genetic program of the multicellular organism.
3. *Tissue and organ level.* This level is caused by specific features of functioning of a cellular set characterized by the same type of organization and certain functions. Such sets of cells form separate tissues, and the cells of various tissues form the organs of a live being.
4. *Organism level.* A separate organism, i.e. an individual, is the elementary unit of this level. It consists of systems of organs (sublevels), each of which performs specific functions to provide the vital functions of the whole organism as an integral biological system. The variety of organisms on the Earth causes the necessity of their classification with the main objective to objectively associate representatives of the organic world into groups (taxa) on the basis of analysis of various signs and features. Two types of such a classification, artificial and natural, are known. The former one is underlain, as a rule, by the results of visual observation; it applies to solving specific tasks (classification by habitat, coloring, movement ability, etc.).

Such an approach is often used when ordering live beings by fossil remains. The natural classification is more scientifically justified; it can have either *phenotypic* or *phylogenetic* character depending on the criterion put in its basis. The phenotypic classification includes data on the morphological, cytological and biochemical similarities but ignores evolutionary relations between organisms. The phylogenetic classification is used to describe the evolutionary relations of organisms; it is underlain by distinctions in the origin and sign inheritance by organisms in the course of their evolution. The live beings belonging to the same taxon are believed to have a common ancestor, i.e. the similarity of their features is caused by a relation. Graphically, it is possible to express these dependences by a genealogical tree called a *cladogram*. At the same time, there is a certain hierarchy (subordination) existing among all known taxa.

5. *Population and specific level*. This level unites populations and species. Populations are sets of live beings of a common look, commonly living and possessing a common gene pool. The species is a set of the populations capable of forming posterities possessing common morphological and physiological signs and relations with the environment. The elementary processes of animate nature evolution begin and proceed just at this level.
6. *Biocenotic and biogeocenotic (or ecosystematic) level*. The biocenosis is the basis of this level, i.e. a set of organisms of several populations (microbiocenosis, phytocenoses, and zoocenosis) united by homogeneous physicochemical conditions (an ecological niche). The main characteristics of a biocenosis are its biodiversity, the density of populations, biomass, productivity, etc. In a biocenosis, the processes of substance circulation and energy transformation proceed as a result of food relations between different organisms. A biogeocenosis (or an ecosystem) is a set of a biocenosis and the site of terrestrial surface (land or reservoir) occupied by this biocenosis.
7. *Biome level*. The level of animate nature uniting the ecosystems located in the same natural and landscape zone (tundra, taiga, desert, steppe, jungle).
8. *Biospheric level*. This is a global ecosystem being a set of biomes and that part of the Earth where life exists. TGeological and biotic substance circulations and energy transformation proceed at this level. Vladimir Vernadsky is considered the founder of the science about the biosphere; he has established the major stability factors of the biosphere to be the heterogeneity of properties of its parts and of substance circulation.

Each of the listed levels of the live matter organization is specific, has its own regularities and, at the same time, depends on the regularities characteristic of other levels.

Some Techniques of Exploring Cells

Differential centrifugation. The distinction of the organelles in a cell by density is the basis of this technique. At very fast rotation in a special ultracentrifuge, the organelles from finely crushed cells sediment from the formed suspension, locating in several layers depending on their density: the denser a component, the lower the rotation speed to sediment it. The layers of organelles thus obtained are separated and analyzed.

Labeled atoms. This technique is effective to study the biochemical processes proceeding inside a cell. A radioactive label is introduced into the substance whose transformation kinetics is under study, i.e. one of the atoms in a substance molecule is replaced by the corresponding radioactive isotope. The ^{3}H, ^{14}C, and ^{32}P isotopes are most often used for this purpose. By observing the radiation from the radioactive isotope inside cells, it is possible to trace transformations in the substance (which this isotope is included into), to establish the sequence of stages of chemical transformations, their duration, the dependence on external conditions, and other characteristics.

Fluorescent microscopy. This technique is based on observation of live cells in UV light. Under the influence of UV radiation, many components of cells start to shine. Some of them exhibit this property naturally, while others require additional influences, e.g., addition of special dyes. By means of this technique it is possible to reveal the sites of concentration of fats, nucleic acids, vitamins, and other substances containing in the cell.

Specifics of Cell Relations in an Organism

All cells in an organism are interconnected. There are some main mechanisms of their interaction.

1. The cells in tissues directly adjoining to each other are connected with channels to provide fast and intense moving of substances.
2. Cells are capable of signal exchange by means of the synthesis and propagation of certain substances. Substances of "near" and "distant" action are distinguished. The former ones restrict their action in the directly adjacent parts of the tissue. The latter ones can reach most remote cells in an organism, influencing their functions. Hormones, the products of the glands of internal secretion are examples of them. To a specific hormone react those cells in whose plasmatic membrane there are the corresponding receptors capable to bind with these hormone molecules. In turn, hormone synthesis is regulated by signals abount the organism state, arriving from all receptors into the interbrain and, further, to the central gland of internal secretion (hypophysis). Such a system is known in vertebrates and advanced invertebrates (mollusks, arthropods, etc.).
3. A special position in the system of cell interaction is occupied by nervous cells (neurons). They coordinate the activity of all parts of the body and provide the organism's interaction with the environment. Nervous cells transmit signals to each other in two ways, namely, by direct electric interaction through special cellular contacts and by means of the so-called mediators, i.e. special substances secreted by nervous and receptor cells to intermediate signal transmission.

The mechanisms of cell interaction provide control of the intensity of protein synthesis and of the quantitative indices of cellular division. Cell fission occurs quicker or slowly in the cells of different tissues. While ontogenesis, many cells gradually specialize and cease to divide. In an adult organism, cell fission occurs only in few updating tissues, e.g., in the hematopoietic, epidermal and lymphoid ones.

6.3. Animate Nature Evolution Concept

Traditional biology is considered to be one of the first stages of biology development. Its object of description was animate nature, and the relevant scientific knowledge had descriptive character (identification and classification of various species of plants and animals). The initial classification (whose basic principles are valid till now) was created in the mid-18th century in the works by Carl Linnaeus (1707–1778); in it, plants and animals were joined in groups by one specific sign. Then, Michel Adanson (1727–1806) introduced the *binary nomenclature* to characterize the genus and species of any live being. Further, the classification was continued by the similarity of the maximum number of signs, whose logic consequence were the principles of genealogical relationship and continuity of the origin of different species of flora and fauna. The importance of this biological knowledge development stage consists in that, first, the foundation has been laid for transition of biological science at a higher level of cognition of molecular and supramolecular structures and, second, these results are the basis for understanding the unity and integrity of the world of live beings (this being one of the main concepts of modern biology).

Evolutionary biology is the most important section of modern biology, whose basis has been put by Jean-Baptiste Pierre Antoine Lamarck's (1744–1829) works, then, it has been developed in the works by Charles Darwin (1809–1882), and, in the 20th century, in molecular and genetic biology researches. The *variability* of this or that species has been shown to be the basis of the emergence of new signs and features in the properties, structure and functions of organisms.

The universality and uniqueness of this factor is in the absolute dissimilarity of live organisms of even the same species, which, finally, manifests itself in the behavior of live beings in the environment. Two types of variability are met in nature. The defined (or individual) variability is caused by transfer of signs and features via the heredity mechanism. The undefined (or group) variability is due to changes in an organism arising under the influence various environmental conditions on it. Another factor of animate nature evolution, *heredity*, fixes both the already existing signs and features, and those acquired in the course of life.

The third major factor of biological evolution is *natural selection* which appeared to be the strongest factor to promote elimination from the environment of those organisms who poorly adapt to the changing (especially adverse) conditions of their existence. A special mechanism of selection has been established in nature, which leads to certain progress in the evolution of live beings. Such progress is caused by the following trends characterizing the concept of natural selection as one of the main motive forces of biological evolution:

- the number of the organisms born in any population or species is always more than that which can survive (for various causes — food, diseases, etc.); nevertheless, the average number of a population or species within a certain ecological niche is almost constant;

- as more individuals are born than can survive, constant competition proceeds for vital conditions (food, habitat, a female or male, etc.);
- the advantage of some organisms over others is caused by the ability of the surviving organisms to adapt to environmental changes, i.e. their adaptable properties;
- the live beings having passed natural selection give posterities possessing the best properties and qualities in comparison with the previous generations.

Variability, heredity, and natural selection are considered at a higher qualitative level in modern evolutionary biology, namely, main attention is paid to the mechanism of action of these factors. E.g., mutations have been established to be major processes to cause changes in a population's gene pool.

These processes, arising either casually or under the influence of external (environmental) factors, can have irreversible character and even be cause of the formation of a new species. Such a situation by no means imply reduction of the role of natural selection since mutation processes in organisms and natural selection are interconnected and interdependent. Besides, unlike Darwin's theory (in which an individual is the main structural unit of animate nature), in modern evolutionary biology (usually called the synthetic theory of evolution) a population is considered to be the main structural unit.

Like natural selection, biological evolution as a long-term process is also characterized by certain trends. In particular, this is a trend of temporary acceleration, which is experimentally proven by data on the origin of organisms of various levels. E.g., the origin of the first live beings (monocellular) occurred about 3.5 billion years ago, multicellular – 2.5 billion years, plants and animals – 400–500 million years, birds and mammals – 100–200 million years, the man – 40–60 thousand years ago.

Another tendency is associated with the evolution of various populations and species proceeding with different rates during different historical periods. E.g., now it proceeds more quickly, due to the influence of features of the civilization processes on the properties of the environment.

The emergence of a new species occurs quicker than the subsequent ones, i.e. the evolution of a population retards in time. Besides, new species are, as a rule, formed from quite simple ones rather than from the most developed ones.

The *concept of hierarchy of the structural levels of animate nature* has a special importance for biological knowledge development. Such a hierarchy of the organization of matter begins with elementary particles, includes atoms, molecules, and supramolecular structures (cells, tissues, biological liquids), continues with organisms, populations, biocenoses and biogeocenoses, and finishes with a global ecological system — the biosphere. The concept of the structural levels of live matter is based on the principles of its systematicity, discreteness and, at the same time, integrity, i.e. every previous level enters into the subsequent one, and the lowest (by development) level is contained in the highest one.

One of the examples of such a hierarchy is a quite distinct correlation between the characteristic sizes of various structures and the periods of their life (Figure 2), namely: the larger their size, the longer they live (atoms, molecules … organisms, populations ecosystems).

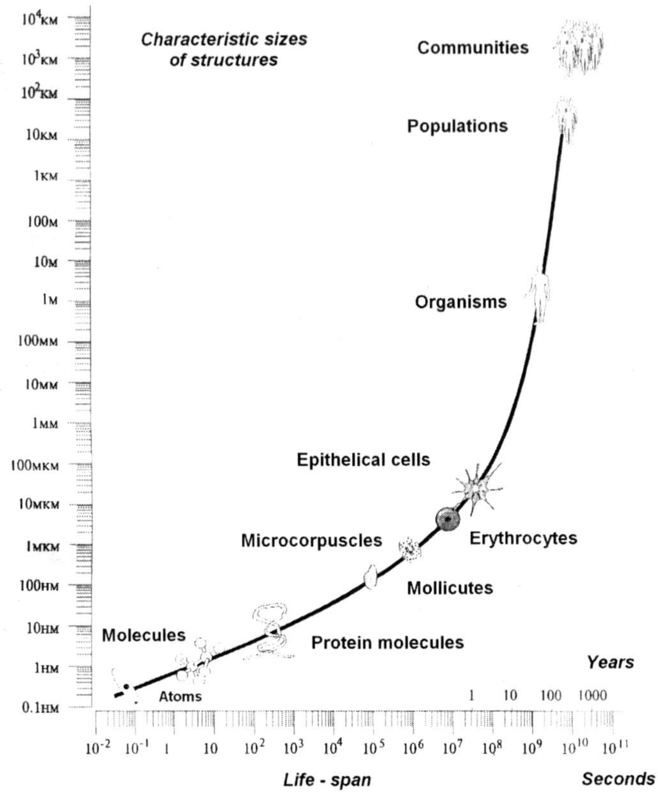

Figure 2. Correlation between the characteristic sizes of the structures of various hierarchies and their lifetime in the biomass. The dependence is expressed in logarithmic scales.

Along with natural selection, biochemical evolution is another motive power of the evolution of live beings, caused by changes in the mechanisms of the chemical reactions proceeding in biological systems. The problem of either casual or natural origin of life on the Earth is closely associated with biochemical evolution. Some researchers consider life emergence as a casual event (e.g., the creationism and panspermia concepts). However, the majority of biologists do not support such a viewpoint. At least, the following data speak in favor of the concept of the natural origin of life. First, the structure and functions of live substance have been established; second, the development of live substance is proven to obey certain physicochemical laws; third, many experimental facts indicate the strictly directed character of nature development. On the basis of these and other data, a conclusion has been drawn that the transition from inorganic to organic compounds (i.e. from the lifeless to the living) consists of three conditional stages:

1) synthesis of initial organic compounds from inorganic ones under the conditions of the primary state of the terrestrial atmosphere and surface;
2) formation of hydrocarbons and biopolymers in primary terrestrial reservoirs from the organic compounds accumulated in them;
3) the most complex process including such stages as:
   - self-organizing of more complex organic compounds;
   - metabolism emergence on their basis;

- reproduction of organic structures of a certain composition;
- formation of an elementary cell.

It has been experimentally established that: i) the physicochemical basis of the listed stages of the final stage of life emergence is the same in all terrestrial organisms; ii) all vital processes occur with direct participation of vegetative and animal polymeric molecules; iii) low-molecular-weight substances (water, mineral salts, simple fats, carbohydrates, etc.) are that medium on whose background, biopolymers perform their functions. Proteins, polysaccharides, and polynucleic acids are most important of them, each of which possesses features characteristic of high-molecular-weight compounds.

The problem of the origin of life on the Earth is considered in more detail in section 6.5.

6.4. Concept of Biopolymers

Many molecular and biological processes are caused by such properties of macromolecules as their chain structure, elasticity, flexibility, etc. The basic role is played by features of the structure and a peculiar hierarchical structure of biopolymers. Four levels of the structural organization of macromolecules are generally characteristic. The *primary structure* is a sequence of various units in the chain, formed at biosynthesis, including cross-links inside chains and between them.

The *secondary structure* is defined by the type of spatial arrangement of the elements in a chain. Usually, it is a so-called near order which is either an ordered helical arrangement of units (α helices) or a formation looking as small folds (β sheets). Besides, the biopolymeric chain as a whole can have a three-dimensional, or *tertiary structure* defined by a distant order (a globule) or a disorder (a coil) in the arrangement of its units. The secondary and tertiary structures are caused by interactions of the units whose character is generally associated with the existence of charges in them. E.g., in curtailed globules, their non-polar protein groups and the lateral groups of the polar remains aspire to appear inside or outside the molecule, respectively. Sometimes, the *quaternary structure,* under which is understood the arrangement of separate links of the tertiary structure inside the complex consisting of several chains, has importance. E.g., hemoglobin possesses such a structure, consisting of four protein macromolecules, each of which a *heme* is bonded with (an iron-porphyrin complex).

Proteins. Protein macromolecules (or polypeptides) are formed in the reaction of joint polymerization of twenty amino acids. All proteins are same in a chemical relation. The macromolecule of any protein consists of identical peptidic groups, to each of which a lateral radical is attached to form, together with the peptidic group, an amino acid residue:

$$\begin{array}{ccccccccccc} & & & & O & & & & & & O \\ & & & & \| & & & & & & \| \\ -NH & - & CH & - & C & - & NH & - & CH & - & C- \\ & & | & & & & & & | & & \\ & & R_1 & & & & & & R_2 & & \end{array}$$

The general structural formula of α amino acids is:

$$
\begin{array}{c}
\quad NH_2 \\
\quad | \\
R - CH - COOH
\end{array}
$$

The set of the residues of these amino acids and the order of their arrangement define the *primary structure* of a protein. Amino acids are linked with each other by a covalent (peptidic) bond (C–N). Its formation occurs at the expense of the amino group ($-NH_2$) of one amino acid and the carboxyl group (–COOH) of another one to release a water molecule. Any live cell contains thousands various proteins, each of which performs some functions peculiar to it only. The peptidic bond is rather rigid and its conformational mobility is limited. However, in each amino acid unit there is an ά-carbon atom providing the presence of two unary bonds round which rotation is possible in this unit. The full amino acid sequence in proteins is usually described as the reduced designations of amino acids (Table 2). The values of the molecular weight of proteins are usually within a range from 10^3 to 10^5. This entails the variety of protein properties which is caused by both distinctions in their composition and the variety of their structure since (as was already told) a huge number of combinations of configurations are probable in one macromolecule. The hormone insulin is the lowest-molecular-weight protein consisting of 51 amino acid residues. For a polymeric protein molecule constructed of twenty various amino acids, with a molecular weight of 2,500, the number of possible isomers (configurations) is about 10^{18}, and for a protein with a molecular weight of 120,000 the number of isomers reaches 10^{1268}.

One of the main features of protein molecules is their ability to spontaneously coil into spatial structures usually called secondary and tertiary ones. The *secondary structure* of a protein macromolecule is more complex: i) one part of the amino acid residues is arranged more or less ordered relative to each other, as a helix; ii) their other part is located in disorder but uniting among themselves in the form of a coil. Both parts (helical and coiled) alternate naturally. E.g., depending on the protein nature (in muscles, blood, milk or other supramolecular structures) the prevalence of either helical or coiled chains is possible.

That is why in most cases such a combination of secondary structures forming the *tertiary structure* of a protein is observed, in which either the fibrous or globular structure prevails. Fibrous proteins generally perform mechanical and construction functions (in hair, nails, feathers, etc.).

Globular proteins, many of which are in a dissolved state in the organism, have much more functions (hemoglobin, insulin, etc.). The stability of the tertiary structure of protein is caused by the existence of a large number of H-bonds between peptidic groups and the functional groups of the amino acid residues. Besides, the experimental data obtained in recent years specify that the macromolecules, including several polypeptide chains (substructures) not linked among themselves covalently, have a *quaternary* structure, which, unlike the secondary and tertiary ones, is characterized by an exacter stoichiometry. Proteins are now classified by such factors as the shape of their macromolecules, physical properties, biochemical functions, solubility (generally in water and weak salt solutions), and molecular structure. The majority of proteins possess hydrophility and form colloidal solutions (the size of particles 4–100 nm). By measuring the intensity of light scattering from such a colloidal solution, it is possible to calculate the molecular mass of the pure protein.

To estimate the molecular mass of a protein is also possible by measuring the osmotic pressure of its solution:

$$PV = nRT,$$

where n is the number of mols of the protein in the volume V; p the osmotic pressure; R the gas constant; T absolute temperature.

For a weight m this expression looks as

$$M = mRT / pV.$$

Ultracentrifugation gives the most reliable results of molecular weight determination for proteins.

Table 2.

α-Amino Acids Found in Proteins

Name	Abbreviation	Structural formula*
Amino acids with nonpolar side chains		
Glycine	Gly (G)	$H{-}\underset{}{C}H(\overset{+}{N}H_3)CO_2^-$
Alanine	Ala (A)	$CH_3{-}CH(\overset{+}{N}H_3)CO_2^-$
Valine†	Val (V)	$(CH_3)_2CH{-}CH(\overset{+}{N}H_3)CO_2^-$
Leucine†	Leu (L)	$(CH_3)_2CHCH_2{-}CH(\overset{+}{N}H_3)CO_2^-$
Isoleucine†	Ile (I)	$CH_3CH_2CH(CH_3){-}CH(\overset{+}{N}H_3)CO_2^-$
Methionine†	Met (M)	$CH_3SCH_2CH_2{-}CH(\overset{+}{N}H_3)CO_2^-$
Proline	Pro (P)	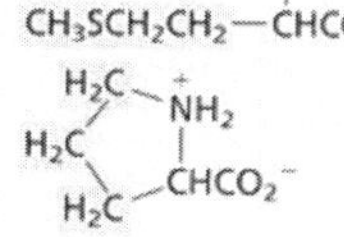
Phenylalanine†	Phe (F)	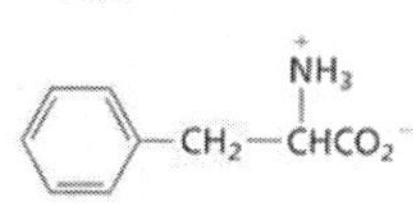
Tryptophan†	Trp (W)	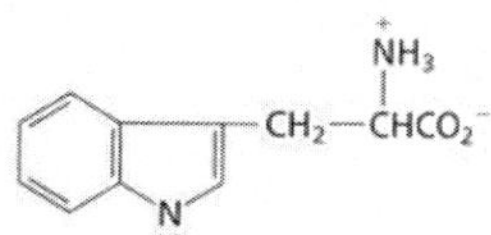

Table 2. (Continued)

Amino acids with polar but nonionized side chains		
Asparagine	Asn (N)	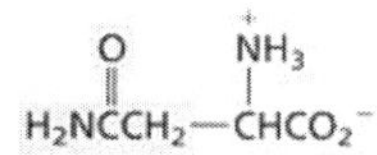
Glutamine	Gln (Q)	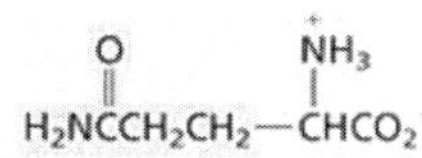
Serine	Ser (S)	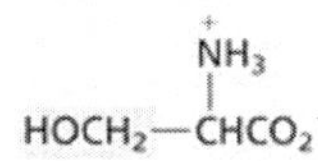
Threonine[†]	Thr (T)	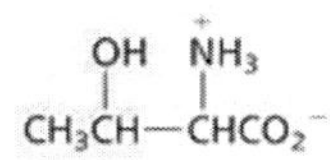
Amino acids with acidic side chains		
Aspartic acid	Asp (D)	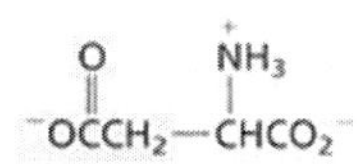
Glutamic acid	Glu (E)	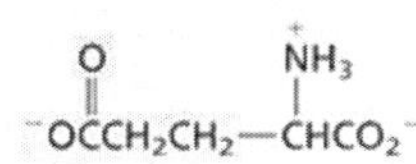
Tyrosine	Tyr (Y)	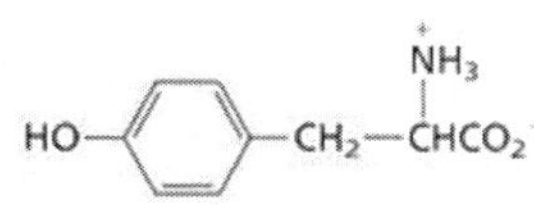
Cysteine	Cys (C)	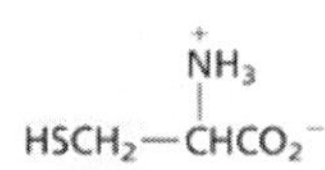
Amino acids with basic side chains		
Lysine[†]	Lys (K)	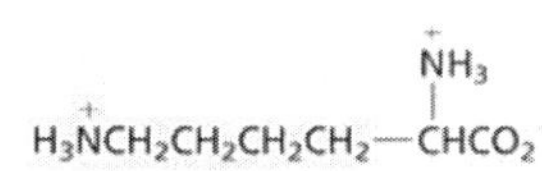
Arginine[†]	Arg (R)	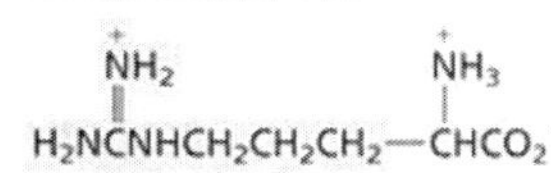
Histidine[†]	His (H)	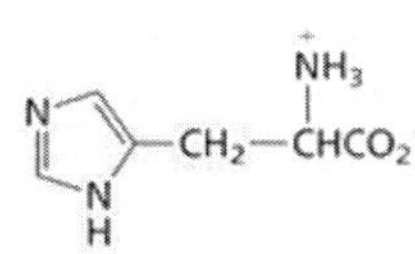

[*]All amino acids are shown in the form present in greatest concentration at pH 7.
[†]An essential amino acid, which must be present in the diet of animals to ensure normal growth.

By structure, proteins can be simple and conjugated. The simple proteins are polypeptides (albumins, globulins, histones, protamines, glutelins, prolamins, scleroproteins…) whose complete hydrolysis produces amino acids only. The conjugated proteins are such as glycoproteins, phosphoproteins, chromoproteins, lipoproteins, whose hydrolysis, except free amino acids, gives substances of other nature.

X-ray analysis, spin labeling, electronic paramagnetic resonance, and nuclear magnetic resonance are most widespread techniques to explore the structure of proteins; their usage has allowed deciphering the majority of natural proteins.

The acidic and basic properties of proteins are mainly determined by the lateral radicals of the amino acids capable of ionization. The dissociation constants of the lateral radicals of free amino acids somewhat differ from those in proteins since the ionization degree of the groups in proteins depends on the nature of neighboring groups, i.e. on the electrostatic environment. The majority of natural proteins belong to acidic proteins, so they carry, as a whole, a negative charge at pH close to neutral.

Proteins actively participate in almost all vital processes proceeding in organisms. The most important functions of proteins are *catalytic* (biocatalysts of enzymatic reactions), *nutritious* (sources of N, P, S, L-ά-amino acids, vitamins and microelements), *transport* (transfer of oxygen, carbon dioxide, fat acids and vitamins, metal ions and amino acids, hormones and the final products of metabolism), *protective* (e.g., in the blood fibrillation system, antibodies of the immune system, the system of protein antioxidants, a body's integumentary tissues), *contractive* (muscular tissues), *structural* (cellular elements), and *hormonal*. All this allows drawing a conclusion that protein macromolecules are responsible for exchange processes in the organism. The main purpose of metabolism consists in the maintenance of a certain level of orderliness in the organism and its parts. This task, on the one hand, is solved by processing of certain substances in the organism, which leads to synthesis of compounds necessary for vital functions.

On the other hand, metabolism removes everything not necessary for the organism from it and what cannot be reprocessed and reused inside. Therefore, metabolism includes a large number of biochemical reactions leading to the formation and decomposition of protein bodies. All this occurs strictly directedly, economically, and reliably.

One more aspect of the protein classification is division of all amino acids into replaceable and irreplaceable (Table 5). The irreplaceable amino acids must surely be part of the protein macromolecules of the man's diet, but, unlike the replaceable ones, they cannot be formed in the organism of animals and the man. If there are all the irreplaceable amino acids in their necessary quantities in the consumed food protein and all of them are completely acquired on digestion (hydrolysis) in the gastroenteric path, the biological value of such a diet is conditionally accepted for 100, while a diet lacking, at least, one irreplaceable amino acid, has no biological value at all. The real minimum of human protein consumption almost twice exceeds its theoretical value.

Table 3. Classification of amino acids

Irreplaceable	Replaceable
Arginine	Alanine
Valine	Asparagine
Histidine	Asparagic acid
Isolcucinc	Glycinc
Leucine	Glutamine
Lysine	Glutamine acid
Methionine	Proline
Threonin	Serine
Tryptophan	Tyrosine
Phenylalanine	Cysteine

Polysaccharides. The macromolecules of these biopolymers are built of monosaccharides with a heterocyclic structure, in which one of the atoms is oxygen, and the others are carbons:

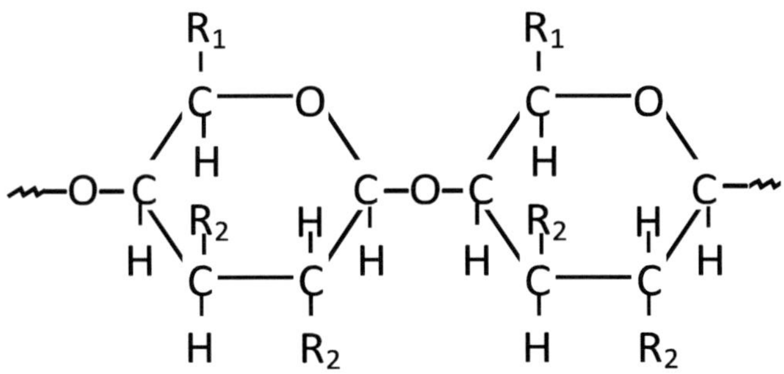

They also possess many physicochemical properties of common polymers, their molecular mass can vary within quite wide limits (reaching several hundred thousands); they are either water-insoluble or form colloidal solutions with water as a gel. Starch, cellulose, glycogen, and some other substances of vegetative and animal origin are typical representatives of natural polysaccharides. In particular, *starch* is a mixture of a linear polysaccharide (amylose) and a branched one (amylopectin) which are characterized by different physicochemical properties. Starch is composed of a mixture of polysaccharides of different structures: amylose – a linear polysaccharide, amylopectin – a branched polysaccharide and the structure of branching parts in the polysaccharide chain.

Glycogen is very close to amylopectin by structure (it is sometimes called animal starch), but with considerably more branched macromolecules. *Cellulose* is the polysaccharide most widespread in nature:

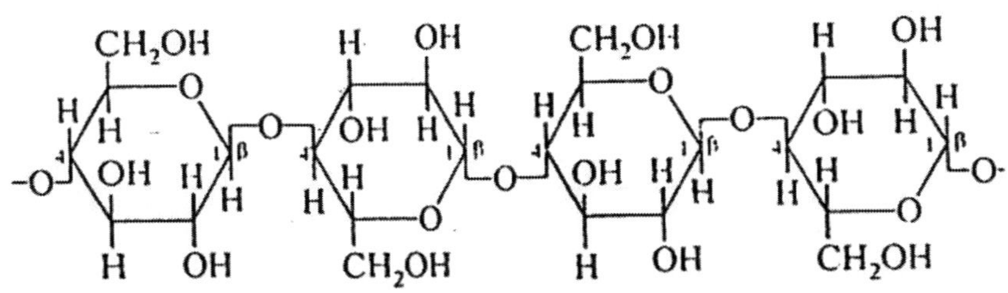

It is called *cell*ulose as a component of the cellular walls in higher plants. Cellulose is a necessary component of the human diet since it actively influences the functioning of the gastroenteric path (peristaltic), stimulates the release of digestive juice, clears the intestines from slag, and reduces the risk of tumoral diseases of the thick intestine. The main properties of cellulose distinguishing it from other polysaccharides are due to the specifics of the configuration of its polymeric chain which is more favorable for the emergence of intermolecular H-bonds.

Polynucleic acids (or polynucleotides). The *deoxyribonucleic* (DNA) and *ribonucleic* (RNA) acids are main of them taking active part in protein synthesis and hereditary information transfer.

Their monomeric unit is a nucleotide consisting of three fragments, namely, the residue of a phosphoric acid molecule, the residue of a sugar molecule (deoxyribose or ribose), and a nitrogenous basis: P–S–B. Most important in the DNA macromolecule are four nitrogenous bases: two pyrimidin ones – thymine (T) and cytosine (C), and two purin ones – adenine (A) and guanine (G); RNA chains have a similar structure differing from DNA only by the sugar type (ribose) in the main chain and thymine replaced by uracil (U):

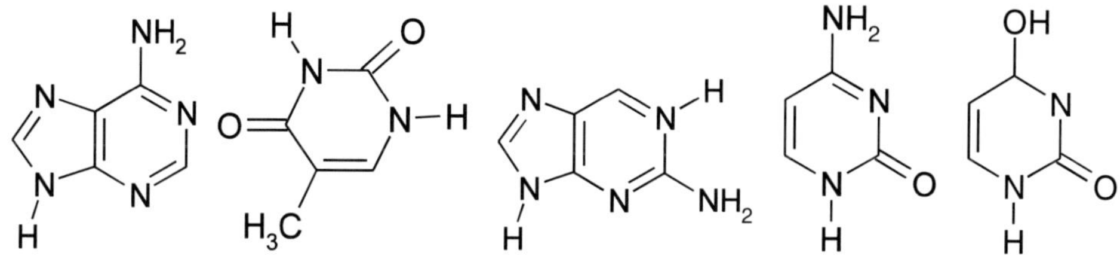

(A) (T) (Г) (Ц) (У)

Francis Crick (1916–2004) and James D. Watson (b. 1928) established the DNA structure according to which its macromolecule represents two helix-like chains connected with two pairs of nitrogenous bases (the secondary DNA structure). The diameter of each helix is about 2 nm. One helical turn comprises 10 nucleotide units; such a structure repeats every 3.46 nm along the helical axis, i.e. 0.346 nm is a share of one unit of the chain. The paired character of the nitrogenous bases in the DNA molecule is caused by their ability to form H-bonds, which gives durability to the secondary structure. This means that for two polynucleotide chains to be associated, they must be complementary: thymine in one chain *vs.* adenine in the other one, and cytosine *vs.* guanine.

Such specific H-bonds make the helices interdependent, and each of the two chains is as though a mould of the other. Therefore, the reduplication principle of DNA molecules underlies Crick—Watson's model, which promotes autoreproduction and reproduction of molecular and, then, cellular genetic structures (Figures. 3 and 4).

The number of nucleotides in a DNA molecule reaches several tens and hundreds thousand. The order of their arrangement (so, the pairs of nitrogenous bases) can be most various. The combination of nucleotides and a various sequence of base pairs along the whole length of the double helical chain create a qualitative distinction of these macromolecules, which underlies the specifics of the hereditary information stored in DNA. Just because of the huge number of various combinations of nucleotides, separate organisms of even the same species are unlike.

Despite the helical shape, the molecules of DNA are quite rigid structures; if such a macromolecule possessing a molecular weight about 10^7 is stretched to the greatest possible degree as a cylindrical spiral rod, its length would be about 5 μm.

Actually, the average size of DNA molecules is less by 1–2 orders of magnitude at least, since the double helix is curtailed into a coil; its internal structure depends on the surrounding physicochemical conditions (temperature, solvent nature, environment acidity, etc.).

The determination of the nucleotide structure of DNA in various live beings has shown that in any tissue of any organism the quantitative (molar) content of adenine is precisely equal to that of thymine, and the cytosine : guanine ratio is 1 as well, i.e. [A] = [T] or [A]/ [T] = 1, and [C] = [G] or [C] /[G] = 1.

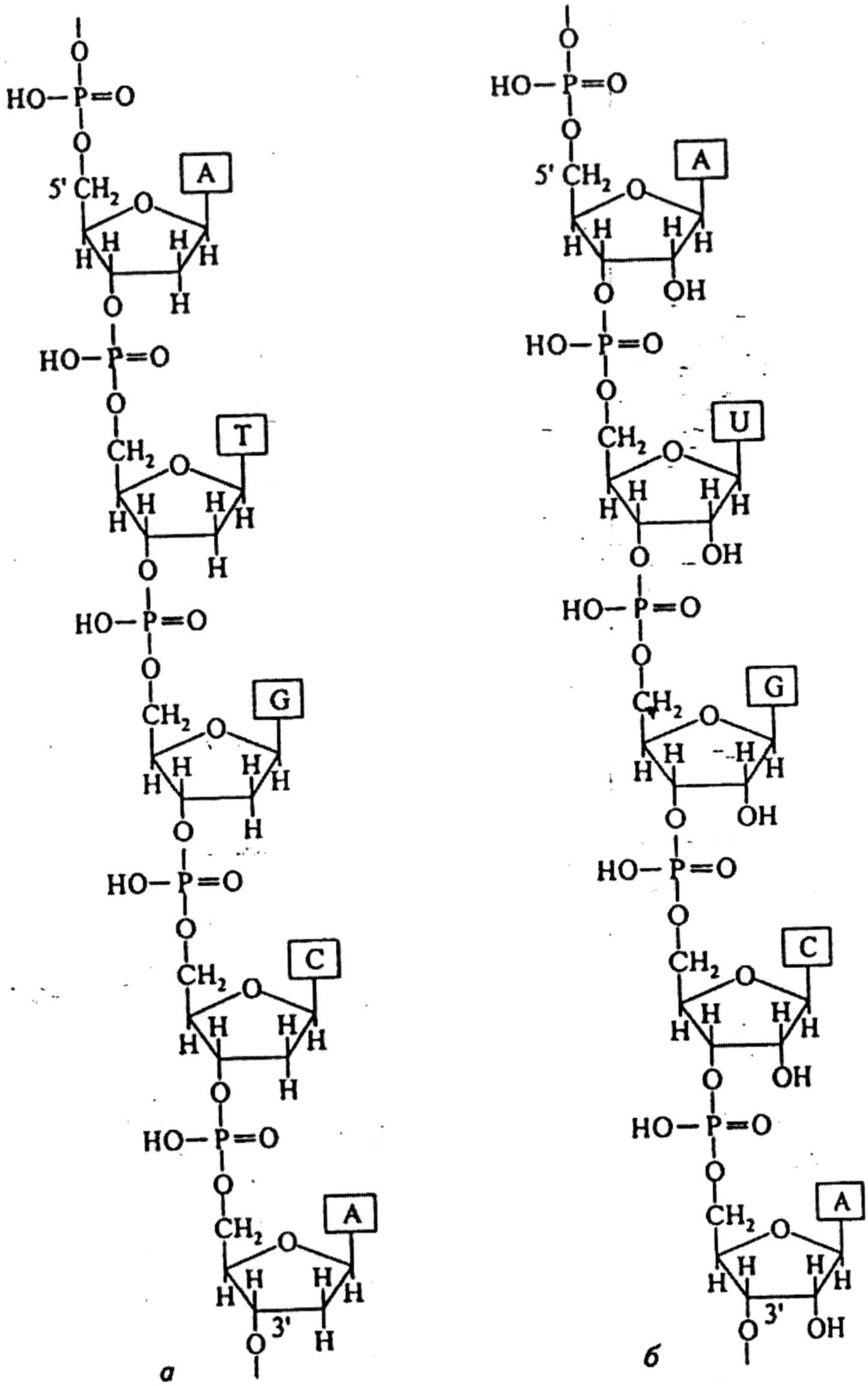

Figure 3. Fragments of primary structures: *a* – DNA and *b* – RNA. In both cases, the phosphate links between nucleoside residues are formed with participation of 3'- and 5'-hydroxylic groups of the pentose cycles (the framed letters denote nitrogenous bases, respectively, Adenine, Thymine, Guanine, Cytosine, Uracil).

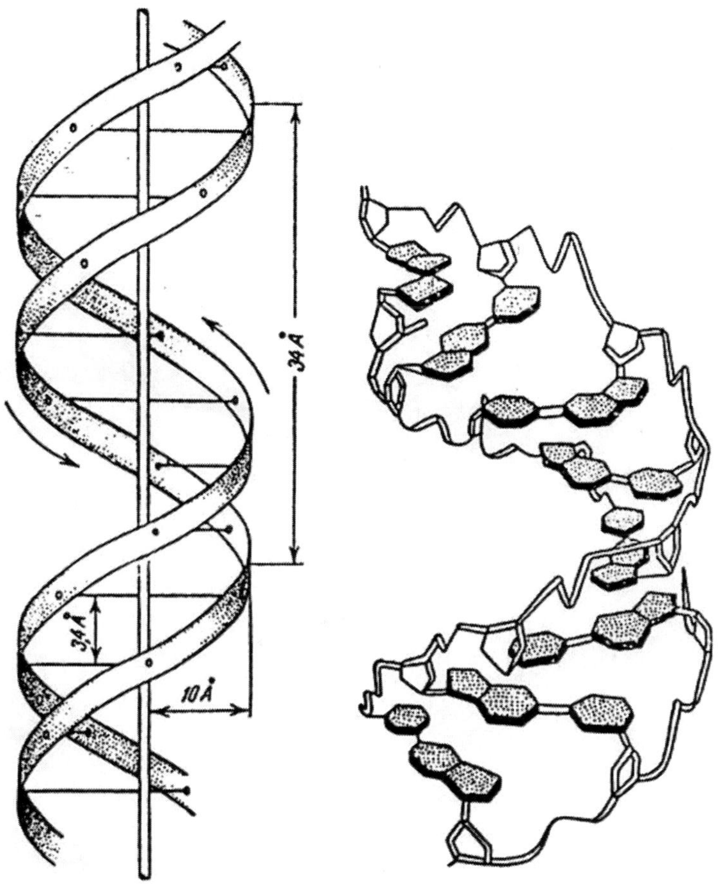

Figure 4. A scheme of the DNA molecule structure from two helically-twisted polynucleotide chains (*to the left*) and a structural model of DNA (*to the right*). Flat heterocyclic bases are located inside the helix as a pile, one over another.

These ratios are called Erwin Chargaff's rules. Whence, ([A] + [G]) / ([T] + [C]) = 1, i.e. the sum of the purin bases (irrespective of their origin) in any DNA macromolecule is equal to the sum of the pyrimidin bases. This allows one to rearrange [A] and [T] and to write:

([G] + [T]) / ([A] + [C]) = 1.

The same concerns RNA macromolecules as well, with the only difference of uracil instead of thymine.

From the written expressions it is also possible to derive:

([G] / [A]) = ([G] + [C]) / ([A] + [T]).

This ratio is called the specificity factor and is accepted as a comparison unit. E.g., in microorganisms the specificity factor varies within quite wide limits, from 0.45 to 2.8. In higher organisms (plants, animals, and the man) the range of its variation is much narrower: 0.55–0.95. E.g., in wheat it is 0.94; in the man 0.66.

After the establishment of the DNA macromolecular structure, there appeared the possibility to influence hereditary signs. A new scientific and practical area in biology, genetic engineering, emerged. As some of its achievements it is possible to present such results as producing necessary medicines (insulin, interferon, growth hormones, etc.), isolation of DNA molecules from the bone tissue (e.g., in forensic medical examinations), changing some hereditary features in a number of large pets, etc.

Besides, a distinctive feature of the DNA macromolecule is its ability to self-reproduction, i.e. construction of a similar molecule near itself. The mechanism of this process can be schematically presented as follows. At the first stage, the double DNA helix is split into two nucleotide chains by breaking the H-bonds between the nitrogenous bases. At the second stage, each of the two separate chains attracts free nucleotides which are always present in the cellular nucleus. At the third stage, the molecules of these nucleotides are arranged in a strict order relative to each separate helical chain (i.e. pairwise: adenine–thymine, guanine–cytosine). At the fourth stage, these nucleotides bind with each other by polymerization, forming a new "daughter" chain. At the fifth stage, the "daughter" chain of nucleotides unites with the "mother" one by formation of H-bonds between the nitrogenous bases to form a new DNA macromolecule. The structure of the DNA molecule and its sites (genes and genomes) as material carriers of hereditary information is reproduced just by such a mechanism, which causes its transfer from cell to cell, and, finally, from one generation to another one.

Modern physicochemical techniques have allowed us to establish that:

- the chemical structure of the DNA molecule (the structure of nucleotides and the existence of nitrogenous bases) is essentially identical in all organisms
- the average sizes of the DNA molecules almost do not change in the organisms of the same species;
- the size of the DNA macromolecules is not identical in various live beings (microorganisms, plants, animal);
- a variety of nucleotide sequences is characteristic for genes and their sets (genomes) as well as for the molecular DNA chain;
- the human genome contains specific nucleotide sequences to cause the uniqueness of properties of the man in comparison with other live beings; just by means of such superchangeable sites of the genomes, called mini-satellites, it was possible to get as if a "portrait" of the human DNA (sometimes they are called "genetic fingerprinting" or "DNA profiling");
- a change in a sign or a property in an organism can also occur without a gene mutation since there is a possibility of changing the spatial arrangement of a gene in a chromosome;
- there are special mechanisms of genome stabilization and even restoration of its damaged sites as a result of self-organizing processes.

As to RNA, this polynucleotide has some features in comparison with DNA. First, RNA is a single-chain macromolecule. Second, it is less stable to the influence of various internal and external factors (temperature, medium pH, etc.). Third, there are several types of RNA macromolecules by their structure and properties, they doing several functions. E.g., the messenger RNA (mRNA) receives information from DNA of what amino acids and in what sequence must enter into the polypeptide chain of protein. The transfer RNA (tRNA) delivers amino acids to the place of protein synthesis (to a ribosome). The ribosomal RNA (rRNA) takes direct part in the polycondensation process of amino acids. Fourth, the quantity of RNA in an organism depends on the type of live beings, the conditions of their activity, some features of the development of the organism, and the cell type. In particular, the maximum quantity of RNA molecules is contained in such cells which vital substances are formed much more intensely in.

Thus, the reproduction system in a live organism performs its main functions by means of polynucleic acids: RNA as though calculates the genetic information stored in DNA, transfers it to the medium containing the initial compounds for protein synthesis, and "builds" the necessary protein macromolecules from them. The information of the primary structure of proteins is coded as a genetic code (a sequence of nucleotides and pairs of nitrogenous bases) whose main features are *triplet nature* (a sequence from three bases – a triplet), *degeneracy* (the same amino acid can be coded by two or several triplets), *overlapping* (the same basis can be part of two next triplets), and *universality* (the principle of genetic information record is identical in all organisms). The implementation of the information stored in DNA includes two stages, namely, *transcription* (enzymatic synthesis of RNA molecules at the corresponding DNA sites) and *translation* (protein synthesis from amino acids in ribosomes according to the information coded in mRNA).

6.5. Concepts of Life Origin on the Earth

The problem of life origin on the Earth is extremely complex in many respects. First, the idea of world creation as of "God's creative act" arose; this hypothesis under the name "creationism" underlies all religions. The Bible has borrowed the ideas of world creation from ancient Babylonian and Egyptian myths which are fantasy and mysticism products. The ancient Greek philosophers of the Milesian school (the 8–6th centuries BC) put forward a hypothesis of the emergence of live beings with no intervention of spiritual forces (in particular, from water). The concept of self-generation of live beings was developed in the works of the philosophers Thales of Miletus (624–547 BC), Democritus, and Epicurus (341–270 BC). In their opinion, the emergence of life is a natural process as a result of some natural powers. Aristotle considered that organisms could originate from both other organisms and inanimate matter, and just existence contains the internal purpose of development, which inhales life into matter as the purposeful internal essence. In the mid-17th century, Francesco Redi (1626–1697) made first experiments on self-generation, having proven that white worms (met in meat and fish) were the larvae of flies and there would be simple rotting of products without their access. In 1862, Louis Pasteur (1822–1895) experimentally proved that the sudden emergence (spontaneous self-generation) of microbes in various types of decaying tinctures and extracts is not life emergence but a result of the activity of microorganisms (rotting and fermentation), whose germs were brought from the outside, i.e. the living occurs from the living only.

Denial of the self-generation concept has led to some scientists' promotion of a hypothesis that life has been existing eternally and that "life germs" infinitely wander in space until get to certain conditions. Such representations supported by Helmholtz (1821–1894) and

W. Thomson (Lord Kelvin) (1824–1907) were further developed by Arrhenius (1824–1907) in the *panspermia theory*. In this theory, Arrhenius assumed that bacterial spores and viruses could be carried away under the influence of electrostatic forces from the planet where they existed and move over space under the pressure of light waves radiated by stars. Then, having incorporated with a dust particle, having thus increased its mass and overcome the pressure of light, a spore can get to the vicinities of a next star and to be grasped by other planet. These assumptions allowed him to draw a conclusion that live matter was capable of being transferred from one space system into another one.

The modern ideas of life origin are based on the works by Oparin, Holdeyn, Bernal, et al. which have gained wide recognition. According to them, life emergence on the Earth has been prepared by a long process of matter evolution during hundred millions years. And the possibility of abiogenesis (life emergence from the lifeless) is caused by specific abiotic conditions existed on the Earth. The primary atmosphere almost did not contain oxygen (no more than 0.1% of the modern quantity) which was formed due to photodissociation of water vapor under the influence of UV radiation of the Sun and decontamination of the basalt magma. Owing to lack of oxygen, and, therefore, ozone, UV radiation easily got through the atmosphere, which created favorable conditions for the formation of such organic substances as amino acids and nitrogenous bases from the gases of the ancient atmosphere, whose molecules are part of some proteins, nucleic acids, and other components of live matter. Due to the lack of oxygen, organic substances did not oxidize to carbon dioxide and water but decomposed into separate fragments serving as an initial material for the synthesis of more complex substances. They were gradually accumulated in separate optimum places, which led to the formation of more or less steady condensations, or *coacervates*, which had some border with the surrounding solution. Gradually, they gained the ability to division and self-reproduction, and, then, to selective absorption of the necessary chemical compounds from the surrounding solution and to releasing unnecessary substances. This was the beginning of metabolism. With the emergence of these processes, the background of the development of life (chemical evolution) finished, and the coacervation drop turned into an elementary organism.

In the mid-20th century, abiogenic synthesis of protein-like organic substances was experimentally carried out in the conditions to reproduce those of the early Earth. The most ancient remains of live organisms have been found in the layers of the lithosphere formed about 3.5 billion years ago. Anaerobic bacteria were possibly the first species of live organisms, in which their metabolism proceeded without oxygen participation. First photosynthesizing organisms appeared about 2.5 billion years ago. Oxygen accumulation in the atmosphere, accompanied by carbon dioxide absorption, became the most important result of their activity. However, the increase in the oxygen content was a long time restrained by the existence of free iron in the Earth's mantle. It was extracted from the rift zones of the median oceanic ridges and dispersed over the whole ocean. Being oxidized by microalgae, iron (as its oxides) was deposited in ocean water areas, especially on shoals. In the Earth's geological chronicle, this period is sharply highlighted with these deposits of iron ore formations. And only with the disappearance of free iron (a most powerful absorber of oxygen) from the Earth's mantle, oxygen began to accumulate in the atmosphere in appreciable quantities. This led to the emergence of an area in the atmosphere at heights of 25–40 km with an increased concentration of ozone which molecular oxygen turned into on absorption of rigid UV radiation of the Sun. The ozone layer protected the terrestrial surface from biologically active solar radiation, which allowed live organisms to leave the ocean waters and to get to the land. Traces of the most ancient ground plants are dated by the Ordovician Period. The development of ground vegetation considerably accelerated the biosphere evolution.

The constantly increasing content of oxygen in the atmosphere promoted ammonia oxidation to form nitrogen. In such a way, the nitrogen–oxygen atmosphere of the Earth was gradually created. But the most part of the oxygen released as a result of photosynthesis for the whole history of our planet did not remain in the atmosphere but was buried in the lithosphere as carbonates, sulfates, oxides of iron and other sedimentary formations. Carbon (as carbon dioxide) was buried as well. Deposits of coal, oil, and other combustible minerals became the result of the biochemical activity of live organisms.

In the early Paleozoic Era, organisms with a solid skeleton not met before were fast settling in new places. Fishes, amphibians, reptiles, insects appeared. The blossoming of reptiles, large reptiles (dinosaurs) occurred in the early Mesozoic Era (Triassic Period). First mammals then appeared.

Dinosaurs died out in the late Cretaceous. The life on the Earth gained further development in the Cenozoic Era, when the vegetative and animal worlds gradually became close to the modern ones. In the early Quaternary Period, a sharp climate change occurred, characterized by a general cold snap and the periodic emergence of extensive continental glaciers (Pleistocene) in middle latitudes, after whose end the modern, post-glacial era (Holocene) began. The Quaternary Period is also called the Anthropogenic Period since first people appeared on the Earth at its beginning.

6.6. Thermodynamics of the Development of Biological Systems

The total number of live organisms in every biocenosis, the rate of their development and reproduction depend on the quantity of energy arriving into the ecological system, and the intensity of circulation of chemical elements and their compounds in it, which can be reproduced and used many times.

Substance circulations and energy transformation in nature obey the fundamental laws of thermodynamics whose physicochemical essence was considered above. At the same time, live organisms are open non-equilibrium macrosystems with gradients of the concentration of chemical substances, temperature, pressure, and electric potential.

The possibility of spontaneous emergence of ordered dissipative structures owing to cooperative movement of large groups of molecules is shown in the works by Prigogine, Glansdorff et al., who have created the bases of nonlinear non-equilibrium thermodynamics. Such self-organizing processes underlie a new interdisciplinary scientific lead called synergetics.

Whence it follows that the emergence of a certain order from disordered systems leads to a seeming contradiction between the second law of thermodynamics and the biological ordering of biological systems (entropy reduction). However, as was already told, if open systems are expanded by adding the environment to them, the second law of thermodynamics becomes applicable to such a continuum.

So, to prove that life in the environment does not contradict the second law of thermodynamics, it is necessary to proceed from that: i) entropy has a thermodynamic and (or) statistical sense only:

$$S_2 - S_1 = \int(dQ/T)_{rev};\ S = k\ \ln W,$$

where the subscript *rev* indicates the reversible nature of the process; dQ is the change of the heat quantity; T temperature; k Boltzmann's constant; W the number of the microstates realizing the current macrostate of the system; and ii) according to the second law of thermodynamics, any evolving system aspires to an equilibrium state, in which all its physical parameters accept constant values; the state of equilibrium is reached when the entropy of the "system–environment" set becomes maximum.

In recent years, achievements of the physics of biopolymers have promoted the development of the latest representations of the possibility to study biological systems and their evolution by means of classical thermodynamic; techniques (Danielle, Denbigh, Mitchell, etc.). In Russia, the representations related with the thermodynamic theory of live system evolution are developed by Gladyshev. Such a theory gives answers to the questions: why evolution proceeds? Where it is directed to? What is its motive power?

By now, a rather large amount of experimental data is accumulated to specify that phylogenesis, ontogenesis, and biological evolution as a whole have a strongly pronounced thermodynamic orientation. In a qualitative relation, the sense of evolution of live beings can be understood from the viewpoint of thermodynamics within the *structure formation model* (structures of highest hierarchies emerge from those of lowest hierarchies). According to this model, every natural hierarchical structure has the energy of its formation, average lifetime (relaxation), biomass volume, etc.

This allows studying live systems as quasi-closed ones in some approximation. The common material at building of hierarchical structural levels in the course of biological evolution is chemical substance. Therefore, the physical theory of evolution of live beings is based on analysis and generalization of experimental data connected with transformations of separate chemical substances in the processes of transition from a simple structure to a more complex one.

The most noteworthy facts consist in the following.

1) During their development, such biosystems as an organella, a cell, an organism's tissue, the biomass of a population, etc., are enriched with power-rich chemical substances to force out water. Therefore, the kinetic features of an organism's aging correlate with the changes of water and fat in it. The same concerns the water content in the brain of a live being (depending on the extent of its relative development). Despite the growth of the power consumption of the biosystem, its chemical stability decreases, and the thermodynamic stability of the supramolecular structures of the body tissues increases. It is also necessary to note that the quantity of water, organic and inorganic substances in biotissues are directly related with the values of the Gibbs specific functions G_{im} and G_{ch} of formation of supramolecular and chemical structures, respectively.
2) Data on the changes of the temperature of melting (denaturation) of biotissues in the processes of ontogenesis and phylogenesis are known.

3) Weber—Fechner's law relates the reactions of biological systems to external influences of the environment (light, noise, ionizing radiation, mechanical tension, psychostresses, etc.).
4) A correlation between the characteristic sizes of structures of various hierarchies and the average duration of their life in a biomass has been experimentally established: lower-hierarchy structures live much shorter than higher-hierarchy ones of in biosystems (Figure 3). As was shown above, atoms, common molecules, macro-molecules, microbodies, cells, tissues ... an organism, population etc. are meant under hierarchical structures. This, in essence, means that the evolution of lower-hierarchy structures proceeds much quicker than that of the organism itself, whose biomass serves the environment for its atoms, molecules, cells, etc. Just in this connection, for every biosystem it is possible to isolate its habitation medium, i.e. a peculiar thermostat with certain parameters. Such quasi-closed subsystems possessing their own thermostats, uniting, form a thermodynamic system of a higher hierarchical level.

The main idea of the thermodynamic model of evolution of live beings, uniting all the listed facts, consists in that the Gibbs specific function of formation of a structure (ΔG) serves a quantitative measure of all changes (including the thermodynamic stability of the chemical, supramolecular and other structures). From the offered model, the existence of a correlation follows between the Gibbs specific function of formation of chemical substances from atoms and molecules and the specific Gibbs function of formation of supramolecular structures from these substances. Therefore, the changes in G at formation of any supramolecular structure, including a population, serve as a measure of the extent of its evolutionary transformation.

Therefore, the life expectancy of biostructures (τ), arising on aggregation of chemical substances, supramolecular and other formations, is determined by the change of the Gibbs function $\overline{G}_{im}$ (P = const, T = const) at formation of these structures:

$$\tau \sim \exp(\Delta\overline{G}_{im} / RT),$$

where the symbol "–" means the specific character of G, the subscript *im* belongs to intermolecular interactions. According to the model,

$$\Delta G_{im} \rightarrow min.$$

Generalizing the foregoing, it is possible to present the model of biological evolution as given immediately by Gladyshev:

> "Under the influence of solar energy, thermodynamically stable substances convert into various products of photosynthesis, which, as a result of spontaneous "dark reactions" and according to the laws of chemical thermodynamics, are transformed into various substances. Further, under the laws of the so-called local thermodynamics of supramolecular processes (owing to the aspiration of the Gibbs specific function of biostructures to its minimum), more stable suprastructures are selected, which are accumulated in microvolumes and macrovolumes of systems. First of all, nucleic acids are selected, whose structure and composition (because of the action of thermodynamic

> factors) slowly adapt to the environment, including the nature of proteins, whose structure is determined by DNA itself. This explains the feedback existence (having a thermodynamic basis as well) between the structures of proteins and DNA. Concurrently to synthesis processes, the disintegration processes of chemical compounds proceed. However, live systems resist to it and aspire to preserve their state, which also has the thermodynamic nature, i.e. biosystems reproduce perishing supramolecular structures, and thermodynamics promotes selecting the most stable of them. So, it is thermodynamically favorable for macromolecular chains to couple with similar chains and to surround itself with the updated "young" substance of live organisms. Evolution selects those thermodynamically preferable ways of processes which promote cell fission and DNA preservation. All this proceeds on the background of parameter fluctuations of the environment (thermostats), which, together with other factors, provides life maintenance".

The physical essence of the model of evolution of biological systems can be understood on an example of the interconnected processes of formation (polymerization) and destruction (depolymerization) of such a known polymer as poly(acrylonitrile). If one analyzes the whole sequence of the processes including heterogeneous formation of solid poly(acrylonitrile) (accompanied by precipitation), its depolymerization (under the influence of light), polymerization of the again formed monomer etc., then, in a thermodynamic relation, these processes (of course, in a certain approximation) model the development of an organism and its evolution.

The proposed thermodynamic theory of biological evolution has a number of very interesting and important consequences for both science development and human life. One of them concerns the predefiniteness of the functions of DNA macromolecules in the emergence and development of live beings. If one assumes the thermodynamic orientation of life evolution, three conditions are necessary for DNA function realization, namely: i) the existence of an aqueous medium, ii) the chemical stability of AT and CG pairs, iii) surrounding of the DNA double helix with thermodynamically stabile supramolecular structures. Therefore, the structure of the double helix in comparison with other supramolecular formations features an increased thermodynamic stability (expressed by the value of G_{im}), which causes the accuracy of replication and the high resistance to mutations. Just this "thermodynamic conservatism" of DNA determines the role of this biopolymer as the keeper of genetic information. Another major consequence of applied character consists in the possibility not only to estimate physiological age by means of Gibbs's function but also to regulate it. The average life expectancy of an organism is determined by its hereditary signs and physicochemical and social conditions of the environment. Every instant of life (age) corresponds to the state of the organism characterized by a certain G_{im} value. Therefore, any change in the conditions of the environment leads to a change in the Gibbs specific function upwards or downwards.

It is clear that at reduction of the value of Gibbs' function (due to the transition of the biological system from one state to another one or, from the viewpoint of thermodynamics, from one thermostat to another one), an increase in the average life expectancy should follow.

The possibility of regulation of the physiological age of an organism is due to the correlation between the chemical and supramolecular components of the Gibbs specific function of formation of food biomass revealed by Georgy Gladyshev. In particular, if food possesses a lower caloric content, the value of Gibbs' specific function becomes more and more positive (or less negative). On the other hand, at an identical caloric content of food, depending on the nature of the consumed biomass, various effects of "rejuvenescence" are reached. The gerontological action of some medicines can be explained (at a quantitative level) similarly. Thus, it is possible to make some conclusions of both fundamental and practical character on the basis of the modern concepts of the thermodynamics and macrokinetics of live matter.

1. Life in the Universe (including Earth) arises and develops in full accordance with the universal laws of nature, in particular, the preservation laws, the second law of thermodynamics, and the law of temporary hierarchies.
2. All biological systems obey *the substance stability principle*, whose sense is that any structural unit of live matter (an atom, a molecule, ..., a cell, ... a population ...) is potentially thermodynamically limited in its interaction with both the structures of its "own" hierarchy and the structures of adjacent hierarchies. One of this principle's consequences is the inverse relationship between the stability of chemical bonds in molecules and the stability of the corresponding intermolecular bonds (characteristic of supramolecular structures).
3. The evolution processes of animate nature, consisting of quasi-closed systems close to equilibrium, can be studied by the methods of phenomenological thermodynamics. This is confirmed by many experimental data which point to that live systems can participate in evolution not only under the influence of expansion work but also under the influence of such types of work (energy) as chemical, electromagnetic, gravitational, osmotic, etc.
4. Special "thermodynamically focused" diets promote reduction of the aging rate of live beings, including the man, and improvement of the quality of their life. *The anti-gerontological indicator* of any foodstuff (of either vegetative or animal origin) is caused by its chemical composition and the supramolecular structure contained therein, and also the ontogenetic and phylogenetic age of this product and the source of its origin (features of the environment). The quantitative characteristic of *the gerontological healthiness* of this or that foodstuff is, first of all, the minimum value of the specific Gibbs function of formation of its supramolecular structure.
5. The basic recommendations on dietary nutrition, directed to reduction of the risk of oncological diseases and their treatment, consist in the following:
    - a sharp decrease in the use of sugar, milk, and dairy products;
    - an increased quantity of green vegetables and some fruits of the bean family (soy, haricot, peas, etc.);
    - certain restriction of the consumption of meat with preference to be given to cold water fish species (herring, tuna, cod, trout, salmon, etc.);
    - of the highest gerontological and anti-cancer value are some food oils with a low hardening temperature whose value depends on the age and habitat medium of plants (minus 15°C – minus 30°C) (hempseed, linseed, sea-buckthorn, cedar, sunflower-seed, soy-bean oils);
    - a sharp (threefold, fivefold, and even tenfold) increase in the consumption of the daily dose of vitamins C, E, PP (e.g., nicotinamide), A, the B group, zinc and selenium.

REFERENCES

Eigen M.. Self-Organization of Matter and the Evolution of Biological Macromolecules. *Die Naturwissenschaften*, Springer-Verlag, Berlin, Heidelberg, New York, 1971.

Gladyshev G.P.. Gerontology and Physicochemical Dietology. *Adv. Gerontol.* 2004. Vol. 13. P. 70–80. [Russ].

Gladyshev G.P.. Life is Inalienable Component of Matter Evolution / *Adv. Gerontol.* 2005. Vol. 16. P. 21–29. [Russ].

Gladyshev G.P.. Macrothermodynamics of Biological Evolution: Aging of Living Beings. *Int. J. of Mod. Phys. B.* 2004. Vol. 18, No 6. P. 801–825.

Gladyshev G.P.. On the Principle of Substance Stability and Thermodynamic Feedback in Hierarchic System of Bioworld. *Biology Bulletin.* 2002. Vol. 29, No 1. P. 1–4.

Gladyshev G.P.. *Supramolecular Thermodynamics as a Key to Comprehension of the Phenomenon of Life.* What is Life from the Physicochemist's View. Moscow, 2003. [Russ].

Gladyshev G.P.. *Thermodynamic Evolution Theory of Living Beings.* Moscow, 1996. [Russ].

Gladyshev G.P.. Thermodynamic Self-organization as a Mechanism of Hierarchical Structures Formation of Biological Matter. *Progress in Reaction Kinetics and Mechanism* (An International Review Journal. UK, USA). 2003. Vol. 28, No 2. P. 157–188.

Gladyshev G.P.. *Thermodynamics and Macrokinetics of Natural and Hierarchical Processes.* Moscow, 1988. [Russ].

Gladyshev G.P.. Thermodynamics of Biological Evolution and Aging. *Electron. J. Math. Phys. Sci.* 2002. Sem. 2. P. 1–15.

Goldfein M.D., A.V. Ivanov, A.N. Malikov. *Concepts of Modern Natural Sciences.* Moscow, 2009. [Russ].

Goldfein M.D., L.I. Karnaukhova. *Principles of the Physics of Synthetic and Natural Macromolecules.* Saratov, 1998. [Russ].

Gorshkov V.G.. *Physical and Biological Bases of Life Stability.* Moscow, 1995. [Russ].

Kushner V.P.. *Biopolymers.* Moscow, 1965. [Russ].

Molecules and Cells / Ed. G.M. Frank. Moscow, 1966. [Russ].

Oparin A.I.. *Life, Its Nature, Origin, and Development.* Moscow, 1968. [Russ].

Ponnamperuma C.. *The Origins of Life.* Dutton, New York, 1972.

Vernadsky V.I.. *Biosphere and Noosphere.* Moscow, 1989. [Russ].

Vernadsky V.I.. *Live Substance and Biosphere.* Moscow, 1994. [Russ].

Vernadsky V.I.. *Origin and Eternity of Life.* Moscow, 1989. [Russ].

Volkenshtein M.V.. *Biophysics.* Moscow, 1988. [Russ].

Volkenshtein M.V.. *Physics of Enzymes.* Moscow, 1967. [Russ].

Chapter 7

SCIENTIFIC PROBLEMS OF COSMOLOGY

Each molecule, therefore, throughout the Universe, bears impressed on it the stamp of a metric system as distinctly as does the metre of the Archives at Paris, or the double royal cubit of the Temple of Karnac.

J. C. Maxwell

7.1. ORIGIN OF THE UNIVERSE AND ITS STRUCTURE

The *Universe* represents the whole environment of the material world available to observation. Various types of objects differing in sizes and masses are contained into it, such as elementary particles, atoms and molecules, substances in various states of aggregation of matter, planets, stars, galaxies, and physical fields (gravitational, electromagnetic, etc.). In spite of the fact that the processes proceeding in the Universe cannot be researched in terrestrial conditions, the usual methodology accepted in natural sciences is used for their investigation. This is caused by that the Universe as a whole obeys the same universal laws which describe the behavior of its separate components. The section of physics and astronomy to study the Universe as a whole is called as *cosmology*. Gravitation has major importance for cosmological processes. Defore Newton (Ptolemy, Copernicus, Bruno, Galilei, and Kepler) and after the creation of classical mechanics the Universe was considered as infinite. Now it is accepted that the Universe implies distances of 10^{28} m and time intervals of 10^{18} s of the order of magnitude (3×10^{14} h, 10^{13} days, 3×10^{10} years).

The initial stage of the Universe formation (in essence, its expansion) when the substance density, radiation energy, and temperature were infinitely high (the singular state) is usually called the *Big Bang*. This is the most developed concept with its heart containing the results of the works by Einstein, Friedman (1888–1925), Hubble (1889–1953), Gamow (1904–1968), and Schmidt (b. 1967).

The Universe formation process as a result of the Big Bang is schematically seen as follows. At the most initial moment of explosion, the Universe's size was almost zero, and it itself was infinitely hot. In the process of expansion (the diameter of the Universe increased by 10^{50} for every 10^{-32} s) the radiation temperature sharply decreased and reached about 10 billion degrees in few seconds. Such fast expansion, subsequently called inflation, corresponds to an explosive process. At that time, the Universe consisted of electrons, protons, neutrons, neutrinos, and their antiparticles. Its further expansion and temperature drop was accompanied by an increase in the electron—positron pair annihilation rate, which led to the formation of photons and to sharp reduction of the quantity of electrons. Approx. in hundred seconds after the Big Bang, the temperature decreased down to 1 billion degrees at which protons and neutrons started their merging to form deuterium nuclei. One fraction of deuterium nuclei then turned into hydrogen nuclei, and the other fraction captured protons and neutrons, turning into helium nuclei (containing two protons and two neutrons), and into heavier elements (lithium and beryllium) in small amounts. In few hours after the Big Bang, the formation of hydrogen, helium, and other elements stopped, but the Universe expansion still proceeded during ca. one million years, without being accompanied by special changes in the chemical composition. At last, when the temperature decreased down to few thousand degrees, the electrons and nuclei began to unite to form atoms. *The model of the hot Universe* was experimentally confirmed by the discovery of the relic radiation in 1965, i.e. the microwave background radiation with a temperature of ca. 3 K, and by the existence of high concentrations of helium (everywhere) and deuterium in interstellar space. The Universe cooling, changes in its density, and the emergence of gravitational attraction between its separate areas l terminated its expansion and caused compression. Such processes proceeded long enough and approx. in billion years after the Big Bang, as a result of substance cooling, its compression and fluctuations of its density, the formation of galaxies to rather uniformly fill the whole space of the Universe began.

The development of modern cosmology actually begins with the creation of the special (at first) and general (subsequently) theories of relativity. It was Einstein who offered *the first relativistic model of the Universe*, proceeding from the classical ideas of its stationary nature, i.e. the invariance of properties in time. According to Einstein's hypothesis, the Universe is spatially finite but has no distinctly expressed borders. Naturally, such a concept is abstract, and it was subjected to criticism by Friedman who developed the bases of the non-stationary relativistic cosmology (1922). Proceeding from the possibility of changes in the curvature radius of the world space in time, he showed that the solutions of the equations of the general theory of relativity allowed considering the probability of three models of the Universe. In two of them the curvature radius of space monotonously grows, and the Universe infinitely expands (in one model – from a point, in another one – starting from some finite volume). The third model concerns the probability of existence of the pulsing Universe whose radius of curvature periodically varies. The reality of such models was confirmed by the effect of *redshift* (the galaxies move away far from each other) discovered by Edwin Hubble in 1929. All the galaxies (except for several ones most close to our galaxy) have been shown to go away from each other with certain speeds at expansion of the Universe. These speeds (υ) can be found from the Doppler shift of the spectral lines in the spectra of these galaxies, and at distances about 10 Mpc = 3×10^{23} m they obey Hubble's law: $\upsilon = H{\cdot}R$. The value of H depends on time only, and now its value (Hubble's constant) is within (50–100) km/s × Mpc = (1.6–3.2) $\times10^{-18}$ s^{-1}). Calculations of these speeds allowed estimating the age of the Universe (counted from the singular state) as about 15 billion years. This value gives evidence of the finite nature of time passed from the instant of the Big Bang, which leads to the conclusion about the existence of *the cosmological horizon (Observable universe)*. This notion represents the border to separate the area of space observable at present from the area which cannot be observed now. However, as light from the farther parts of the Universe reaches the observer, the cosmological horizon extends. Now, it is about 6,000 Mpc, i.e. covers more than a half of the space volume of the Universe available to observation.

Analysis of the density of the substance which the Universe consists of has shown that its value sharply falls upon transition from small objects to big ones: it is 10^{14} g/cm^3 in atomic nuclei and in neutron stars, about 1 g/cm^3 in planets and many stars, about 10^{-24} g/cm^3 in the Galaxy, and over the whole observable Universe the average density of substance is close to the so-called critical density estimated as ~10^{-29} g/cm^3.

It is necessary to note that the concept of the non-stationary Universe does not contradict the concept about its uniformity. The uniformity of the Universe manifests itself, first, in essentially identical structural elements of stars and galaxies, the same physical laws and physical parameters, and, second, in the homogeneous distribution of substance over the whole Universe. This allows drawing a basic conclusion that our planet Earth does not occupy any exclusive position in the Universe.

The *Galaxies* are congestions of many stars and stellar remnants; they differ in size and appearance. They are conditionally divided into several types, namely: elliptical, spiral, irregular, interacting, radio galaxies, etc. The *elliptialc galaxies* are ellipsoids with different extents of compression; they are simplest by structure; the density of their stars monotonously decreases from the center.

The *spiral galaxies* are considered as most numerous; our Galaxy (the Milky Way) and the Andromeda Galaxy belong to them. The *irregular galaxies* have no strict center and their structure features no certain regularities. The Large and Small Magellanic Clouds (which are satellites of our Galaxy and have, in comparison with it, considerably smaller masses and sizes) exemplify them.

The *interacting galaxies* are usually located at small distances from each other, and some of them are overlapped. The *radio galaxies* possess very powerful radio emission; one of the known radio galaxies is in Cygnus, and another one is in the Centaurus constellation.

Their radio emission has various causes. In particular, it is assumed that radio-waves are emitted by the ionized hot gas (generally, hydrogen) in interstellar space which is formed by hot stars and space rays. In recent years, it has been revealed that all galaxies are distributed non-uniformly and there are places in the Universe where they are absent at all.

Notes:

1) 1 pc (parsec) is the unit of length equal to the distance at which the parallax is 1 s, which corresponds to 3.26 light years or 3.9×10^{13} km.
2) Parallax is the visible change of the location of a subject owing to moving of the observer's eye.
3) Light year (l. y.) is the way which light passes for one calendar year: 9.5×10^{12} km.

Our Galaxy includes about 150 billion stars, being a huge disk which, in turn, consists of a core and several spiral branches. The diameter of our Galaxy is 120,000 light years, its thickness is 10,000 light years; the mass is about 10^{11} of the Sun's mass; the speed of rotation has different values depending on the distance from its center, and no strict dependence exists. Therefore, various sites of the Galaxy have different periods of rotation. The stars and fogs within the Galaxy move in absolutely various directions.

The Galaxy itself rotates round its axis quite slowly, spending 180 million years for a complete turn. The star density in the Galaxy is also non-uniform; e.g., in the field of its core it reaches two thousand stars per parsec, which exceeds the average star density in the Sun vicinity almost by 20 thousand times. Near the Sun, the average distance between stars is 10 million times higher than the average diameter of these stars; therefore, the stars almost never collide.

The following classification of stars is accepted in modern cosmology.

1. The variable stars are those whose shine, energy, and radiation spectrum are constantly changing. It is due to their periodic pulsation, i.e. expansion or compression of the star, accompanied by temperature changes.
2. The double stars are systems of two stars rotating round their common center of masses.
3. The ecliptic-double stars are those whose total shine is constantly changing owing to periodic overlapping by each other.
4. The multiple stars are systems of three and more stars.
5. The red dwarfs are the stars whose diameter is less by 2–3 times, and whose density is higher by 2–3 times than in the Sun, and the luminosity is comparable with the Sun's.
6. The white dwarfs are the stars whose mass is close to the Sun's one, the diameter is 0.01 of its diameter, the density is ~10 t/cm^3, the luminosity is 10^{-4} of the Sun's.
7. The red giants are the stars whose diameter exceeds the diameter of the Sun by hundreds times, the density is lower by thousands times, and the luminosity exceeds that of the Sun by tens and hundreds times.
8. The neutron stars are huge congestions of neutrons; their mass is close to that of the Sun, the diameter is $2 \cdot 10^{-5}$ of the diameter of the Sun, the density is ~10^8 t/cm^3.
9. The pulsars are the stars whose electromagnetic radiation periodically changes over the whole known range of wavelengths (from radio waves to gamma radiation).
10. The quasars are quasi-star (i.e. similar) sources of radio emission; their mass and diameter exceed those of the Sun by tens million times; the energy of radiation exceeds the total energy of radiation of several galaxies.

7.2. Some Characteristics of the Sun and the Solar System Planets

Nicolas Copernicus' creation of the heliocentric system was a major event in our understanding of the Solar system structure, whose basis contains the following statements:

- the Sun rather than Earth is located at the center of the world;
- the spherical Earth rotates round its axis, and this causes the seeming daily motion of all stars;
- Earth, as well as all other planets, rotates round the Sun, and this explains the apparent motion of the Sun relative to stars.

Now the Solar system is known to include the Sun itself, planets and their satellites, and other space objects (asteroids, comets, meteorites, and even space dust) as well.

According to modern representations, the formation of the Solar system about 5 billion years ago had the following stages:

1) The initial gas-dust cloud reached an appreciable density and started to compress under the influence of gravitational forces.
2) In the process of compression, the sizes of the gas-dust cloud decreased, and the speed of its rotation increased; the speeds of compression of the cloud parallel and perpendicular to its axis of rotation differed, which led to condensation of the cloud and the formation of a disk.
3) At achievement of some limiting density, the dust particles started to collide with each other and the released kinetic energy of the compressing gas-dust cloud led to temperature growth, with the central region of the disk to heat up most intensely.
4) At achievement of a temperature of few thousand degrees, the central region of the disk started to shine; it is a *protostar* (or a *protosun*) which falling of the cloud substance proceeded on, to increase the pressure and temperature at the center.
5) When the temperature at the center of the protostar reached millions degrees, thermonuclear reactions of hydrogen burning began.
6) The external areas of the disk remained relatively cold and, as a result of different cloud areas possessing different gravity, the largest condensations in these external areas formed planets.

The Sun has been established to move almost evenly on a big circle of the celestial sphere called as the *ecliptic*, from the west to the east (oppositely to rotation of the celestial sphere), making a complete turn for ca. 200 million years (a *galactic year*).

Depending on the nature of movement on the celestial sphere, the planets are divided into the lower ones (Mercury, Venus) and the higher ones (all the other planets, except Earth), or the internal and external planets relative to Earth's orbit. In their movement on the celestial sphere, Mercury and Venus never leave far from the Sun (Mercury – not farther than 18–28 degrees, Venus – not farther than 45–48 degrees) and can be either to the east or to the west from it. The instant of the longest angular distance of a planet to the east and to the west from the Sun is called as *the eastern or evening elongation* and *the western or morning elongation,* respectively. The movement of both higher and lower planets from the west to the east is called as forward movement, and that from the east to the west is the reverse one, and these movements alternate by certain cycles.

The question of where exactly the Solar system ends and the interstellar space begins has not still been finally solved, as its border depends on the influence of two various phenomena, namely, the solar wind and solar gravitation. Even outside of the *heliopause*, the Sun can keep other space objects by its gravitation (the heliopause is the border separating interstellar space from the Solar system substance).

The radius of the Sun is 7×10^8 m, the volume 1.4×10^{27} m^3, the surface area 6×10^{18} m^2, the mass 2×10^{30} kg, the average density 1.4 mg/m^3, the density at the center 160 mg/m^3, the pressure at the center 3.4×10^{16} Pa (1 Pa is 10^{-5} atm), the total radiation (power) 3.8×10^{26} J/s, the average surface temperature is about 6,000 K, the temperature at the center ca. 10^7 K, the speed of movement relative to next stars 19.5 km/s, the distance from the Galaxy center about 30,000 l. y., the speed of rotation round the Galaxy center 250 km/s, the cycle time round the Galaxy center 200 million years, the age of the Sun is about 5 billion years. The chemical composition (by mass): ca. 70 % of hydrogen atoms, 27 % of helium atoms, 3 % of other element atoms.

Table 4. Some characteristics of the Solar system planets

Planet	Average distance from the Sun		Rotation Period	Cycle time on the equator	Equatorial diameter, km	Mass (in Earth's mass)	Average substance density, mg/m^3	Number of satellites	Diameter (relative to the Earth)
	mln km	a. u.							
Mercury	57.9	0.39	87.97 days	58.6 days	4 878	0.055	5.7	–	0.38
Venus	108.2	0.72	227.70 days	243 days*	12 104	0.805	4.95	–	0.95
Earth	149.6	1.00	365.26 days	23 h 56 min	12 756	1.00	5.5	1	1.0
Mars	227.9	1.52	686.98 days	24 h 37 min	6 974	0.106	3.94	2	0.53
Jupiter	778.3	5.20	11.86 days	9 h 50 min	142,600	314.03	1.33	12	11.1
Saturn	1,427	9.54	29.46 years	10 h 14 min	120,200	94.01	0.70	16	9.4
Uranium	2,870	19.18	84.01 years	~20 h*	53,000	14.4	1.49	5	4.0
Neptune	4,496	30.06	164.81 years	~20 h	49,500	17.0	2.09	2	3.9
Pluto	5,946	39.75	247.7 years	6.39 days	4,000	0.002	0.4	1	0.47

* The rotation direction is opposite to the revolution direction.

7.3. Cosmological Characteristics of the Moon

The Moon is the sole natural satellite of Earth and the closest celestial body.

Origin of the Moon

Three hypotheses of the Moon's origin are most known. At the end of the 19th century, J. Darwin put forward a hypothesis according to which the Moon and Earth originally constituted one common molten mass whose speed increased in the process of its cooling and compression; as a result, this mass became torn onto two parts: the big one (Earth) and the small one (the Moon). This hypothesis explains the small density of the Moon (formed of the external layers of the initial mass) but meets serious objections from both the viewpoint of the mechanism of this process and owing to essential geochemical distinctions between the terrestrial shell rocks and the lunar ones. According to the "capture" theory offered by Weitzekker, Alfvén, and Urey, the Moon originally was a small planet which, when passing near Earth, due to the terrestrial gravitation, turned into a satellite of Earth. Special calculations have shown that the probability of such an event is very low. According to the third theory developed by Schmidt in the mid-20th century, the Moon and Earth were formed simultaneously by association and consolidation of a large number of small particles. But, as the Moon has, on the average, a smaller density than Earth, the substance of the protoplanetary cloud should have been divided to form a considerably higher concentration of heavy elements in Earth. From whence it follows that Earth surrounded with a powerful atmosphere started to be formed first, and, at subsequent cooling, the substance of this atmosphere condensed into a ring of *planetesimals*, which the Moon was formed of. This theory is most developed scientifically.

Some Physical Characteristics of the Moon

The shape of the Moon is very close to a sphere with a radius of 1,737 km, which is 0.2724 of the equatorial radiuse of Earth. The surface area of the Moon is $3.8{\cdot}10^7$ sq. km, the volume $2.2{\cdot}10^{25}$ cm^3. More detailed estimation of the geometrical shape of the Moon is complicated by the absence of oceans on it and there being no obviously expressed level surface to be a reference for measurements of heights and depths. Besides, as the Moon is always turned to Earth by one side, it is obviously possible to measure the radiuses of points on the visible hemisphere surface of the Moon (except the points on the very edge of the lunar disk) from Earth on the basis of the weak three-dimensional effect caused by *libration* only (see below). Studying of librations has allowed estimating the difference of the principal semiaxes of the Moon ellipsoid. Under the influence of tidal forces, the Moon is a little extended towards Earth. The polar axis is shorter than the equatorial ones directed towards Earth and perpendicular to the direction to Earth, by approx. 700 and 400 m, respectively. The mass of the Moon was more precisely determined from observation from artificial Earth satellites; it is 81 times less than the mass of Earth, i.e. $7.35{\cdot}10^{22}$ kg. The average density of the Moon is 3.34 g/cm^3. The gravity acceleration on the Moon surface is 6 times less than on Earth, i.e. 162.3 cm/s^2 and decreases by 0.187 cm/s^2 when lifting by 1 km.

Movement Features of the Moon

The Moon moves round Earth with an average speed of 1.02 km/s on an approximately elliptic orbit in the same direction in which the vast majority of other bodies of the Solar system do, i.e. counter-clockwise if taking a detached view of the Moon orbit from the North Pole of the world. Owing to its elliptic orbit and perturbations, the distance from the Moon varies between 356,400 and 406,800 km. The Moon cycle time round Earth, the so-called *sidereal* (stellar) month, is 27.32166 days, but is also subject to small variations. Studying of the Moon movement is one of the most difficult problems of celestial mechanics. Elliptic movement represents a certain approximation only since some processes caused by the attraction of the Sun and planets (including the Earth) are imposed on it. The Sun's attraction of the Moon is 2.2 times stronger than Earth's one, so, strictly speaking, one should consider the Moon movement round the Sun and Earth's influence on it. However, of interest is usually the Moon movement as seen from Earth and, consequently, the principal gravitational theory (offered by J. Hill) describes the Moon movement exactly round Earth.

The plane of the Moon orbit is inclined to the ecliptic at an angle of 5°8'43", subjected to small variations. The *ecliptic is* the big circle of the celestial sphere which the visible annual moving of the Sun center proceeds along. The ecliptic plane is inclined to the plane of the celestial equator at an angle of 23°27' and is crossed with the celestial equator at the points of the spring and autumn equinox. *The celestial equator* is the big circle of the celestial sphere representing the line of its crossing by the plane perpendicular to the so-called axis of the

World round which the visible rotation of the celestial sphere proceeds. The points of intersection of the Moon orbit with the ecliptic are called as the ascending and descending knots, have non-uniform movement, and make a complete turn on the ecliptic in 6,794 days (about 18 years). Thereof, the Moon comes back to the same knot through a time interval, a little shorter than the sidereal month (27.21222 days on the average). The periodicity of solar and lunar eclipses is related with this period. Besides, as the Moon rotates round the axis inclined to the ecliptic plane at an angle of 88°28', and with the period precisely equal to a sidereal month, it is always turned to Earth by the same side. The combination of uniform rotation with non-uniform movement along the orbit leads to small periodic deviations from the invariable direction to Earth. So, it is possible to see only 59% and less of the whole lunar surface from Earth at different times; such deviations are called the *libration* of the Moon. The planes of the lunar equator, of the ecliptics, and of the lunar orbit are always crossed in one straight line.

Phases of the Moon

Without being self-shining, the Moon is only visible where either sunshine or the rays reflected by Earth fall. This explains the Moon phases. Every month the Moon, moving along its orbit, passes between Earth and the Sun and is turned to us by its dark side; at this time there is a new moon. In 1–2 days after this, a narrow bright sickle of the young Moon appears on the western part of the sky. The other part of the lunar disk at this time is poorly illuminated by Earth turned to the Moon by its day hemisphere. In 7 days, the Moon departs from the Sun by 90°, the first quarter comes, when exactly half of the lunar disk is illuminated and the terminator (the line between the light and dark sides) becomes a straight line, i.e. a diameter of the lunar disk. In the subsequent days, the terminator becomes convex, the look of the Moon comes nearer to a light circle, and in 14–15 days a full moon comes. On the 22nd day the last quarter is observed. The angular Moon-to-Sun distance decreases, it again gets the shape of a sickle, and in 29.5 days a new moon comes again. The interval between two consecutive new moons is called as the synodic month, whose duration (29.5 days) exceeds a sidereal month, since Earth during this time passes about 1/13 of its orbit and the Moon, to pass between Earth and the Sun again, must additively pass 1/13 of its orbit, which requires some more than two days. If the new moon occurs near one of the corners of the lunar orbit, a solar eclipse is observed, and the full moon near the knot is accompanied by a lunar eclipse. It should be noted that the specified system of the Moon phases was the basis for a number of calendar systems.

Moon's Surface and Its Relief

The Moon surface is quite dark, its *albedo* being 0.073, i.e. it reflects only 7.3% of the light rays of the Sun on the average. The *albedo* is the ratio of the quantity of the radiant energy reflected by a body to the quantity of the energy falling on this body, to characterize the reflective ability of the body's surface. A day on the Moon lasts nearly 1.5 teresstrial days and a night proceeds as much. Without being protected by an atmosphere, the Moon surface heats up to +110ºC in the afternoon, and cools down to –120ºC at night. However, owing to the very weak heat conductivity of the surface layers, such quite high temperature variations penetrate into few decimeters only. Owing to the same cause, the warm surface is quickly cooled during full lunar eclipses. Visually, some irregularly-shaped darkish extended spots are visible on the Moon, which were first mistaken for seas but finally appeared to be plains and ring-shaped mountains (craters). At the end of the 19th – the beginning of the 20th century, a large atlas of the Moon was published according to the photos obtained by the Parisian observatory; later, a photographic album of the Moon was published by the Lick Observatory (the USA), and in the mid-20th century Kuiper compiled some detailed photo atlases of the Moon obtained by means of modern telescopes in various observatories of the world. The relief of the lunar surface was also established as a result of long-term telescopic observations. The "lunar seas" occupying about 40% of the visible Moon surface have, as a rule, a simple cap-like shape of a diameter from 15 to 200 km. The mountain heights have been estimated by the length of their shadows on the surface or photometrically. The compiled hypsometric cards for the most part of the visible Moon side (the 1:1,000,000 scale) are also used to account for the roughness of the edge Moon with the purpose to locate it. The absolute age of the lunar formations (mountains and craters) is precisely known at some points only. However, using some indirect methods, it is possible to establish that the age of the youngest large craters is tens and hundreds millions years, and the majority of large craters appeared 3–4 billion years ago.

Special calculations related with the thermal history of the Moon show that soon after formation its subsoil was warmed up by radioactive heat and considerably melted, which was accompanied by intense volcanic phenomena to form craters and cracks. Besides, at early stages, a very large number of meteorites and asteroids dropped onto the surface of the Moon, at whose explosions craters of different sizes and structures appeared as well. Now, meteorites drop onto the Moon much less often, and the volcanism degree has considerably decreased.

Internal Structure of the Moon

The internal structure of the Moon has been explored as well as Earth's, i.e. by the distribution nature of longitudinal and cross seismic waves. From analysis of the gravitational field of the Moon it has been concluded that its density slightly changes with depth; unlike Earth, no high mass concentration at the center is observed. The upper layer of the visible side of the Moon represents a crust whose thickness is about 60 m. It is very probable that on the reverse side the crust is approx. 1.5 times thicker. In both cases, the crust is made of erupted crystal basalt rocks. However,, the basalts of the continental and sea areas have appreciable differences by their mineralogical composition. The most ancient continental regions of the Moon are mainly formed by a light rock (anorthosite), almost completely consisting of plagioclase minerals with small impurities of pyroxene, olivine, magnetite, titan magnetite, etc. The crystal rocks of the lunar seas, like terrestrial basalts, are generally made of plagioclases and monoclinic pyroxenes (augites) formed when cooling the magmatic melt on the surface or near it. As the lunar basalts are less oxidized than the terrestrial ones, this means that they were crystallized with lower oxygen—metal interaction. Besides, in comparison with the terrestrial rocks, a lower content of some volatile elements and the richness by many refractory elements are observed. Due to the impurity of olivines and, especially, ilmenites, the areas of the seas look darker, and the density of the constituent rocks is higher than on the continents. Under the crust there is the mantle in which, similarly to the terrestrial one, it is possible to resolve the upper, middle, and lower ones. The thickness of the upper mantle is about 250 km, that of the middle one is 500 km, and its border with the lower mantle is located at a depth of ~1,000 km. To this level, the speeds of cross waves are almost constant; this means that the substance of the subsoil is in a solid state, being a powerful and relatively cold lithosphere, in which seismic vibrations do not fade for long. The composition of the upper mantle is assumingly olivine–pyroxenic, and at deeper depths, spinel and the mineral melilite (occuring in metamorphic and igneous rocks) are present. At the border with the lower mantle, the temperature is close to those intervals of temperatures where strong absorption of seismic waves begins. This area is a lunar asthenosphere. At the center there is a small liquid core of a radius less than 350 km which no cross waves pass through. The core may be either iron-sulfide or iron; in the latter case it should be less, which agrees with the estimates of the density distribution by depth. Its mass does not possibly exceed 2% of the total Moon mass. The temperature in the core depends on its composition and varies from 1,300 to 1,900 K.

7.4. Anthropocosmism Concept

The idea of interrelation of the man and the Universe, put forward in an extreme antiquity, has an old tradition in both western and eastern philosophy. For the early representations of anthropocosmism (Tsiolkovsky, Vernadsky, Chizhevsky, et al.) a naturalistic aspect is characteristic. The modern general scientific concept of antropocosmism, being development and generalization of the previous versions, should be constructed on the basis of the sciences of not only space but also the man and his reason. A special contribution can be brought by theoretic and methodological innovations, characteristic of not only the era of cosmos but also the entire period of modernization of such fundamental ideas which initiate civilization development in connection with the problem of transition to sustainable development. In the most general view, anthropocosmism is understood as the world outlook concept to display both the influence of space on the man and impact of the man on space. In other words, anthropocosmism should coincide in many respects with sociocosmism, the concept describing the relations in the society—Universe system. Wide space exploration is predicted as a number of certain stages of extraterrestrial space industrialization. The man's advance into the Universe will be accompanied by development of the production of goods out of Earth, transition from the "two-dimensional" infrastructure of economic activity to a "three-dimensional" one. Just as since the beginning of Neolithic revolution the production of goods became the basis of accelerated social and economic development, now the production activity on Earth and out of its limits becomes the base of the human penetration into cosmos.

At the initial stage of implementation of the ideas of astronautics (cosmonautics—in Russian) its founders and followers dreamed of the man's exit out of the limits of his planet, of massive resettlement of people into the Universe space. E.g., Tsiolkovsky represented the mankind moving into space as a kind of nomads who has left its native planet and, then, the Solar system as well, wandering over the Universe in searching for new sources of power and negentropy. The scientist treated cosmos as that ecological environment, those processes and forces which could accelerate social progress in the conditions of our planet, would eliminate the threat of geological and, subsequently, cosmic disasters.

At the present stage of space exploration, the use of space means for scientific-technical, social-economic and ecological development on our planet appears most important. The concept showing that mankind must be in the focus of cumulative space activity as the process of space exploration has been named as "anthropogeocosmism" or "sociogeocosmism". From this concept it follows that our planet will long be the center of the ecosystem consisting of Earth and its surrounding space. However, further, along with Earth, there will appear other centers of space activity, settlements of people on artificial constructions and natural celestial bodies suitable for life, e.g., on the Moon, Mars, the satellites of the Jupiter group planets.

At the same time, in the anthropocosmism concept, not only space exploration by means of the modern tools of astronomy is of great importance but also solving such problems as determination of the man's place in the system of the Universe and its reason in the Universe, finding a genetic and structural link between the man and space. Concretizing the ideas of our Universe as of manned system and space ecomedium of human development (including the noosphere formation), just the anthropic cosmological principle reveals the relation between the global properties and characteristics of the Universe; first of all, this concerns the relation between the processes of self-organizing and emergence of an hierarchy of structural levels of matter whose evolution has led to the appearance of the man. Besides, the modern anthropocosmism concept is closely linked with the problem of extraterrestrial civilizations.

7.5. Problem of Extraterrestrial Civilizations

The conclusion of the life and reason existence out of our planet is known to be hypothetical. These questions were discussed by the thinkers of antique Greece and New Era. Nowadays the problem of extraterrestrial civilizations is developed by special sciences, first of all, astronomy. A characteristic feature of modern exploring the problem of plurality (habitability) of cosmic worlds, unlike the post-Copernicus period, is the emergence of special scientific works rather than physiophilosophical ones. It is interesting that this research is focused on the problem of communication with extraterrestrial civilizations (this problem is abbreviated as CETI, Communication with ExtraTerrestrial Intelligence). However, approx. since the late 1970s, a new term SETI (Searching for ExtraTerrestrial Intelligence) appeared. The replacement of "communication" by "search" is quite justified, since it is necessary to find extraterrestrial civilizations first.

Since the time of Giordano Bruno, essential changes have occurred in the development of his philosophical ideas of the Universe and mankind. However, the experts engaged in the problem of extraterrestrial civilizations are poorly informed in the field of philosophy and social knowledge. In their search they, as a rule, are guided by a scientistico-technistic approach, paying much attention to the natural-scientific, astronomical conditions of the possible existence of extraterrestrial civilizations and technical aspects of communication with them. The role of economic and ecological factors in civilizational processes in the Universe is almost not considered.

The problem of extraterrestrial civilizations and communication with them is complex, interdisciplinary, and general-scientific. Development of the perspective of extraterrestrial civilizations essentially depends on the possibility of revealing regularities and tendencies of development inherent not only in our terrestrial civilization but also in other expected civilizations in space. During the development of this problem, two main problems are to be solved. The first problem consists in assistance to the progress of those sciences and solving those "terrestrial" problems of the mankind development which proceed from the ideas of our terrestrial civilization as about a systematic and complete developing object transforming into the cosmonoosphere. The second problem is searching for extraterrestrial civilizations, their detection, and contact establishment. If the first problem is urgent just now, the second one appears more fundamental and its solution seems more distant in time. The conceptual character of this problem is caused not only by the basic possibility of contact establishment to other worlds but also by the need of deeper studying the prospects of the survival and development of the terrestrial mankind, revealing the regularities and prospects of noospherogenesis on our planet and beyond its limits. Just this globoterrestrial aspect prevails in the modern research concerning general characteristics and social regularities of the development of expected extraterrestrial civilizations.

Some scientists have an opinion that the solution of the problem of communication with extraterrestrial civilizations can give the chance of using their experience and knowledge. However, the situation is absolutely different: the terrestrial civilization is an information model which the man gets hypothetical knowledge of extraterrestrial civilizations from. Until this knowledge gains a corresponding "space interpretation", it would be fondly to rely on the help of other civilizations. At solving any questions of this problem, the mankind surely proceeds (obviously or implicitly) from the life and civilization in space resembling the terrestrial ones. This is the basic methodological and epistemic-logical principle of every research, and any deviation from it would mean transfer of scientific research to the area of unscientific fantasy. Therefore, it is now the sole effective principle of the probabilistic-theoretical increment of our knowledge of extraterrestrial civilizations. When studying the logic and semiotics problems with the purpose to design a language for interstellar transfers, further enrichment and generalization of sign systems happens. Besides, it has been established that any originally created information means are of substantially anthropomorphous character and expect the communicant having the concepts developed by the people.

The ideas being developed in the works investigating the development regularities of extraterrestrial civilizations (physical, technical, social-economic, ecological, etc.) very much remind the ideas of global problems of the scientific and technical revolution, and especially global modeling. At treatment of the development problem of space civilizations, such quantitative characteristics are analyzed as settling space, the total number of population, the area and volume population density, the total consumed energy, the energy per consumer, the volume of accumulated information, its quantity per consumer etc. The results obtained at the development of global problems, "cleared" from terrestrial specifics and separate tactical nuances, can be used in research of the social aspect of the extraterrestrial civilization problem. The principal breakage with geocentrism, understanding of that the future of our civilization, positive solutions of global problems cannot anymore do without space exploration and the formation of the cosmonoosphere is indicative in this regard. But until extraterrestrial civilizations are found, we can take information from purely theoretical contact to them as hypothetical objects, space copies of our civilization which have made "great transition" to the cosmonoosphere. Analysis of this problem allows us to deepen and expand our ideas of the *noospherogenesis* process at the cosmic development stage. It is a feature of the noospheric-methodological orientation: not only the outlook development of the world's scientific picture but also a certain social efficiency, urgency, the humanistic orientation of the search to comprehensive studying of the man and society.

If one proceeds from the possibility of the existence of extraterrestrial civilizations, each of them cannot be considered as an especially independent civilizational process covered with only general regularities together with other representatives of the social development stage. They can form the basis for a new community in the sense of integrity. It is a question of such association of civilizations into a uniform system (*the astroinfonoosphere*) in which the social stage of matter development will fully show its space nature. The prospective social evolution of matter will represent not only the purely planetary independently proceeding noospheric-civilizational process independent of other social space. Such autonomy, if any, takes place only till a certain historical period of the development of the noospheric civilization exploring cosmos in broad scales. In any case, civilizations of the noospheric development level steadily exist in space, sooner or later they will necessarily come into contact and form an information-noospheric system.

7.6. Dark Matter and Universal Evolution

At the end of the 20th–beginning of the 21st century, scientific discoveries occurred which allow us to look at evolution processes in a new way. Our Universe is supposed not to be lonely and not to any more represent a Universum containing all the real. What was earlier called the Universum is now considered one of *miniuniverses*. Expansion of our representations of the evolution problem and an essentially new concept of universal evolution are substantially caused by the discovery of new and mysterious substances, namely: *dark energy* and *dark matter*. Before the discovery of dark energy, *hidden substance* (or *dark mass*) was ascribed to dark matter. If this second form of matter occupies about 20% of the mass and energy of the Universe, then the whole modern science occurs to study only few percents of the whole material content of the Universe. Therefore, the modern scientific picture of the world must represent a systematic and complete image of both inanimate and animate nature, the man and mankind, obtained on the basis of synthesis and generalization of the whole scientific knowledge. In spite of the fact that since the second half of the 20th century the ideas of a universal evolutionism lie at the heart of the general scientific picture of the world, from time to time this or that science unexpectedly makes a contribution which appears very principal for understanding of the Universe and the human role therein.

7.6.1. Basic Forms of Being As Ways of Matter Self-preservation

As was told above, philosophy has always aspired to investigate the universe in its integrity and at various levels, to reveal the most fundamental and limiting primary origins and life bases therein. At the physiophilosophical stage, it aspired to get into nature's secrets, to find eternal and passing, steady and changeable, universal and specific in it. The interest to these questions began to be lost with the development of separate sciences, and the philosophy of science focused its attention on the philosophical and methodological analysis of topical problems and on the logic aspects of the philosophy—science relationship.

At the end of the 20th century, the so-called dark energy was discovered and active search for experimental (generally observant) confirmations of its existence began. This unusual form of space matter being still poorly blends with the "material and evolutionary" scientific picture of the universe being created recently. Science accepts the hypothesis (though in a debatable form) about the existence of a very stable part of the Universe which, in the generalized look, we will call *dark matter* consisting of two main forms, namely, *dark energy* and *dark mass* (Table 5). The definition of the notion of "dark matter" is ambiguous. Some researchers believe that this notion, in general, is hardly applicable since there is already "dark substance" (dark mass) which, as well as the term "dark energy", is really dark because radiates no light.

They prefer a different name for dark energy, namely, "space vacuum". By different estimates, the dark components constitute 96–97% of the total material content of the Universe. They are invisible unlike the shining (or seen) Universe and remain, to some extent, in a poorly or generally non-evolutionary form against the distinctly expressed evolution of substance. Dark energy resists to gravitation and allows, owing to the "world antigravitation" property, the Universe to extend with acceleration. It is supposed that space vacuum, influencing the Universe expansion (antigravitation), is a rather stable and not changing form of matter influenced by nothing. By modern representations, the property of self-preservation over evolution prevails in this form of matter being. Many scientists assume that space vacuum could have undergone several phase transitions before the event called the Big Bang occurred. Therefore, a certain transformation of vacuum passed through several phases could precede the Big Bang.

So, dark matter, which is unobservable and does not radiate, consists of two parts essentially different in their nature, namely, the anti-gravitating dark energy with a density of about 67–75% of the total energy density of the Universe, and the "hidden substance" whose density is about 23–25% of the total energy density of the Universe, exceeding the density of usual observable substance (i.e. the "shining" Universe) by 5–6 times. On the share of the latter there is only 3–4% of the universal energy density.

Dark mass, being exposed to gravity forces, interacts with neither substance nor radiation; it absorbs nothing and does not shine, however, some change processes distinct from substance evolution are quite distinctly traced therein. The development of gravitational inhomogeneities in dark mass during the period before electron recombination speaks for this. There was also a decrease in the dark mass density due to which universal gravitation ceased to dominate during a certain period, having conceded to the forces of "universal antigravitation".

Probably, "hidden substance" consists of very long-living components of a very high density (from neutron stars, cooled-down white and brown dwarfs, black holes, including relic ones, and other components arisen as long ago as at the Big Bang).

Thus, the modern cosmological (still mainly hypothetical) picture of the world gives us unknown or poorly-studied forms of matter self-preservation which is the cause and source of the existence of stars in a galaxy, congestions of galaxies, super congestions, and other forms of baryonic matter (quarks, bosons, leptons) which constitutes the observable Universe.

Table 5. Fundamental Forms of Matter Being in the Universe

Material (observable) part of the Universe —baryonic matter: 3% of the world energy density;	*Dark energy* (space vacuum): 67% of the world energy density;	*Dark mass* (hidden substance): 30% of the world energy density.
Average density of mass (stars, molecular hydrogen clouds, etc.) 2×10^{-31} g/cm^3	Average density of mass 7×10^{-30} g/cm^3 (identical over the whole universe)	Average density of mass 2×10^{-30} g/cm^3
Characteristics of development 1. Evolves (remains through evolution). 2. Obeys the law of universal gravitation. 3. Expands for about 7 billion years with acceleration	*Characteristics of development* 1. Does not change and does not evolve on the Big Bang, evenly fills the whole universe. 2. Possesses the property of anti-gravitation, causing accelerated expansion of the Universe. 3. The composition and structure are unknown; it is supposed as a homogeneous formation. 4. Dark energy is influenced by neither substance nor dark mass, but a back influence has been established:	*Characteristics of development* 1. Changes but does not evolve as normal substance. 2. Obeys the law of universal gravitation. 3. Consists of very heavy particles of unknown nature which poorly interact among themselves and with usual substance

The nature of self-preservation of large-scale structures of the material world is related with the stability of our Universe during many billions years. This self-preservation as the dominating and fundamental component of matter being assumes some asymmetry between the dark and baryonic forms of matter. From the viewpoint of modern natural sciences it is important to answer the question: how can matter self-preserve without evolution and, the more so, without any changes? To answer now this question is essentially difficult because neither thermodynamics nor synergetic representations allow constructing a model of such matter self-preservation during nearly one and a half tens billions years. The more so that it is the dominating part of the Universe (nearly three quarters of its mass and power contents). If dark energy remains throughout many billions years, either still unknown laws of preservation must act therein or it is necessary to assume the total absence of movement there, characteristic of material matter. This means that, according to modern synergetic representations (applicable to the material part of the Universe), there exists an unusual way of matter preservation in space vacuum (the more so that dark energy is not self-organizing substance).

7.6.2. Dark Matter Influence on Global Evolution

Approx. in the first 300 thousand years after the Big Bang, dark mass generated gravitational heterogeneity of the substance distribution in the Universe which galactic congestions and galaxies were further formed in. Without its influence, global evolution would be impossible, since galaxies could not be formed in the absence or insufficient quantity of dark mass. In the first minutes and hours after the Big Bang, the distribution of common substance in the Universe was uniform and remained such till proton—electron recombination at the age of the Universe of about 270 thousand years. The gravitational condensation of common substance was interfered by the pressure of radiation which this substance intensely interacted with. Dark mass did not interact with radiation and nothing interfered with the formation of gravitational condensation. A certain structure of inhomogeneities was already formed in the Universe at that time, consisting of gravitating dark mass only. After electron recombination, common substance simply fell into the gravitational potential holes prepared by dark mass. If the dark mass did not have time to form "dark" *protogalaxies*, then galaxies could not be further formed from common substance, and this substance would dissipate in the Universe. Moreover, both modern galaxies and their congestions cannot exist out of the potential holes formed by dark mass. It is not excluded that dark mass can interact with common substance not only through gravitation. Interaction of this part of dark matter with common substance through weak interaction (as in the case with neutrino) is also probable.

Until recently, at discussion of the global evolution problems no questions arose concerning the role of dark matter in this process. Generally, it was spoken that some global characteristics, first of all, the main physical constants corresponding to the four types of fundamental interactions, their fine tuning, and some already known parameters of the Universe (the dimension of space and time, topology etc.) are such which allow the evolution process, including the global evolution with the man on its top now. At the same time, the existing dark energy as space vacuum with a constant and not changing energy density makes very essential impact on the processes of evolution of the Universe's material part. Predominating in our Universe, dark energy surpasses three times all the other forms of space matter taken together by energy density, creating powerful world antigravitation. With the Universe age of 6–8 billion years, the era of cosmological expansion with acceleration began because the dark mass density gradually decreased and became lower than the vacuum density. This anti-gravitational expansion of the Universe replaced the cosmological era of gravitation domination over antigravitation and the material forms of matter over vacuum (dark energy).

The question of the world invariance was discussed since Einstein who believed the Universe to be invariable and static. But in 1917, having applied his general relativity theory to cosmology, Einstein unexpectedly found that the cosmological model created by him did not confirm the eternity, invariance, and static character of the Universe.

To save his ideas of the static character and invariance of the Universe, he therefore introduced a so-called cosmological constant as one of the fundamental physical constants. The first cosmological model of the world offered by Einstein represented an ideally symmetric model of the Universe in space and time. Such an idea remained till Edwin Hubble's 1929 discovery of space expansion, the galaxies moving off from each other.

Thus, the idea of evolution and global evolution in the general scientific picture of the Universe, in connection with the discovery of dark sides of the Universe, is essentially transformed and an essentially new world outlook is formed. That type of evolution, which was often called as universal evolution, appears not so universal as considered quite recently.

7.6.3. Cognition Problem of Dark Matter

The cognition problem of the dark forms of matter has its own specific features. The *black hole* is one of such components of dark mass, which is a closed sphere formed as a result of collapse of a massive star and the trapped substance having got into it cannot leave it even as radiation because of enormous compression (gravitational attraction). Hence, the black hole is understood as a "space—time" area where the second space speed is equal to the velocity of light and the gravitational field does not let out even photons. The spatial border of a black hole was named as *the event horizon* outside of which getting no information of any events and conditions in this black hole is possible. Any space body, substance, and radiation are capable to get in, fall into a black hole, but cannot leave it. This means that the cognizing subject cannot get any information of the internal state of a black hole when being out of it. Especially it applies to the central part of a black hole, to singularity as a superdense state where, as is supposed, no known laws of nature are obeyed. However, the space matter making a black hole does not lose the whole information since it can be characterized by the mass, electric charge, and the own moment of momentum. Other characteristics and properties of the matter constituting a black hole can further be found if it really contains information. Therefore, no matter what a black hole has been formed of, inside it the vast majority of the earlier available distinctions disappear in the superdense homogeneous medium ($\sim 10^8$ t/cm^3).

While the interior of a black hole cannot be perceived by the external observer, the existence of this latent space object can be revealed, in particular, owing to the enormous gravitational fields representing powerful potential energy sources which, in principle, can be released when substance gets onto the event horizon of the black hole. At this loss (accretion) of substance on the black hole, very large amount of energy can be radiated before it crosses the border (the event horizon) of the black hole (the existence of an X-ray aura round it is possible). Certainly, if there is no substance and radiation in the vicinity of the black hole, it will remain unfound. If in the vicinity of a black hole there is substance and it falls on the black hole, the effect of radiation emission (as though particles take off from this hidden space object) is simulated for the external observer. At getting of the substance of stars and gas clouds into supermassive black holes, their brightness will appear the most intense in our Universe as large energy quantities (by two orders of magnitude higher than those in nuclear reactions) are released in this case. Such supermassive black holes with most powerful gravitation can serve as "dark" (hidden) power sources in a number of space processes with huge power yield and luminosity (quasars, the cores of active galaxies). A black hole with its mass of about three million solar masses is apparently in the core of our Galaxy.

The cognition problem of dark matter and other hidden exotic superdense space objects is a special gnoseological problem because the reliability of the results of scientific search is estimated from indirect, side signs of their influence on the usual shining matter. However, analogies between the physics of black holes and thermodynamics, and between them and the gravitation theory are used in cosmological modeling. The cognition of space vacuum is also complicated by its interacting with nothing, including the observer, though its influence on substance on the scale of the Universe not simply takes place but is determining for the future universe. The evolution processes and related information processes play an important role not only in the Universe but also in its cognition by the man. The relation between the Universe and its properties has found reflection in the anthropic cosmological principle which is realized in the observable Universe. The main idea of this principle is that the man and the material Universe assume their mutual existence.

REFERENCES

Arkhangelskaya I.V., I.L. Rozental, A.D. Chernin. *Cosmology and Physical Vacuum*. Moscow, 2007. [Russ].

Cherepashchuk A.M., A.D. Chernin. *Universe, Life, Black Holes*. Friazino, 2005. [Russ].

Chernin A.D.. *Cosmology: the Big Bang. Friazino*, 2005. [Russ].

Khvan N.P.. Furious Universe: from the Big Bang to Accelerated Expansions, from Quarks to Superstrings. Moscow, 2006. [Russ].

Nicolson I.. *Gravity: Black Holes and the Universe*. Halsted Press, 1981.

Parshakov E.A.. *Origin and Development of Solar System*. Moscow, 1999. [Russ].

Rubtsov V.V., A.D. Ursul. *Problem of Extraterrestrial Civilizations*. Kishinev, 1987. [Russ].

Ursul A.D., T.A. Ursul. *Evolution, Cosmos, Man*. Kishinev, 1985. [Russ].

Uspensky G.R.. *Cosmonautics of the 21st Century*. Moscow, 1994. [Russ].

Chapter 8

BASIC CONCEPTS IN THE EARTH SCIENCES

Though the documents of Earth are incomplete but they tell the truth only. It is our reading of the remained pages of the geological chronicle which can be erroneous.

S. V. Meyen

8.1. EARTH STRUCTURE, ITS MAIN CHARACTERISTICS AND MODELS

Earth is the third (by distance from the Sun) planet of the Solar system, largest of the four internal planets. Its mass is 6.0×10^{21} t, the average radius 6,371 km. All the internal planets have rather small sizes and high densities due to the content of heavy oxides, silicates, and the existence of a ferruterous core. Increased concentrations of oxygen, nitrogen, carbon, and lack of hydrogen and helium are characteristic of them. The external planets possess considerably lower densitшуы and differ by composition. Some of them, e.g., the giant planets Jupiter and Saturn mainly consist of a hydrogen–helium mixture close to the Sun composition. These distinctions are due to the corresponding chemical differentiation of the protoplanetary gas-dust cloud, whose exact mechanism is not clear yet. The Earth rotation axis forms an angle of 23.5° with a perpendicular to the ecliptic plane, which causes season changes. The first attempts to estimate the size of the Globe were made long time ago (Erathosphen, approx. 3rd century BC) but only at the end of the 17th century the question of Earth's shape and sizes began to be considered from the physical viewpoint as well as the geometrical one. Isaac Newton in his "Mathematical principals of natural philosophy" stated the theory of the Earth shape on the basis of the law of universal gravitation. However, the acceleration of gravity and the shape of our planet depend not only on the force of mutual attraction of a body with it but also on other forces, in particular, the centripetal one caused by Earth rotation. This force (a part of the gravity force) is directed perpendicularly to the rotation axis. The resultant of these two forces in the case of a spherical planet forms an angle different from 90°, with the tangent line to the sphere surface at the point of a body' location. As a result, a body on the surface of a rotating sphere is influenced by a force aspiring to shift it towards the equator. The man does not notice it, because under the influence of the centrifugal force the shape of Earth has changed in such a manner that at every point its surface is perpendicular to the resultant force, i.e. Earth has the shape of an *ellipsoid of rotation* whose polar radius R_p is shorter than its equatorial one R_e. The gravity on the real Earth is equal to the difference of the normal components of the gravity and centripetal force and depends on geographic latitude. At a rather strict approach, the surface of Earth is described by an individual figure called as the *geoid*. The divergence between the surfaces of the geoid and ellipsoid does not exceed several tens meters while the equatorial and polar radii differ more than by 20 km. According to modern measurements, R_e = 6,378.1 km, R_p = 6,356.7 km. The compression of the geoid is $(R_e - R_p) / R_e = 0.34\%$. The average radius is defined as the radius of the sphere having the same volume as the ellipsoid of rotation has.

The found deviation of our planet's shape from the spherical one assumes that the constituent substance possesses some plasticity characteristic of viscous liquid. Space researches have allowed measuring the value of compression more exactly. It has appeared that it does not completely correspond to the balance figure of a rotating liquid planet. Therefore, there is a deviation from the equilibrium state, and there must be tension in the depths of Earth. From whence it is possible to conclude that the material of the terrestrial subsoil possesses durability and, in this sense, is distinct from a liquid.

Information on the mechanical properties of the material of the globe can be obtained from observation over the *tidal deformation* experienced by Earth under the Moon's influence. Every point of the terrestrial surface rises and falls under the influence of the tide-generating force (the force of mutual attraction and the centripetal force). These movements occur very smoothly, but have quite high amplitude (up to 2/3 meters). However, owing to the huge wavelength, insignificant deformations unnoticed by the man thus appear. The nature of tidal movements depends on the properties of substance. Their analysis leads to a conclusion that the substance of Earth as a whole is quite close, by properties, to ideally elastic body though inflows are somewhat late relative to the tide-generating force. This means that the subsoil of Earth possesses some viscosity. The data hitherto known allow considering the substance of Earth as possessing both viscosity and hardness, and it can thus be neither an ideal liquid nor an ideally elastic body. It is accepted to call such materials as viscoelastic which can show the properties of either a solid body or a liquid in various conditions. E.g., from geological observations it follows that at long-term (above 10^4 years) orogenic processes, the rocks involved as though flow similarly to liquids with an average viscosity of 10^{20}–10^{22} P whereas at the influence of not too long acting forces (few seconds–days) they behave as usual solid elastic bodies. Now it has been established that the main processes transforming Earth arise on deep depths having a rather complex structure. Their research is possible by physical methods only. However, it is not always possible to unequivocally describe such a complex natural object as our planet. Therefore, the notion about *the models of Earth* has been introduced to represent some simplified schemes which reflect main lines of its structure.

Seismic Model and Internal Earth Shells

The most complete data on the internal structure of our planet are given by observation of the distribution of elastic *seismic* waves arising within Earth owing to earthquakes or artificial explosions. Two types of body seismic waves are distinguished, namely: the longitudinal P-waves (elastic waves of compression which are oscillations of substance along the wave propagation direction, i.e. along the ray), and transverse S-waves (elastic shift waves to reflect the oscillations of substance across this direction). Both waves propagate in the substance of Earth like light rays in optical media, being reflected and refracted on interfaces where their speed changes depending on the mechanical properties of the medium. The body seismic waves, in a literal sense, expose our planet to a radiation and allow revealing the internal structure of Earth without direct penetration into its subsoil. Such representations were put forward by the Russian geophysicist prince Boris Borisovich Galitzine, one of the founders of seismology, at the early 20th century.

From the theory of elasticity it follows that the speeds of propagation of P-waves (υ_p) and S-waves (υ_s) are determined by the following expressions:

$$\upsilon_p = \sqrt{\frac{K + 4/3\mu}{\rho}}, \qquad \upsilon_s = \sqrt{\frac{\mu}{\rho}},$$

where K is the modulus of dilatation; μ the modulus of rigidity; ρ the substance density.

As is seen from these relationships, longitudinal waves extend faster and so are called the primary ones (P-waves), while transverse waves are the secondary ones (S-waves).

The centers of the majority of earthquakes are located not deeper than 100 km from the Earth surface, but in the case of the largest ones they can be deeper (down to 700 km). At theoretical consideration, the source of seismic waves is represented by a point called the *focus* or *hypocenter*, and its radius projection onto the Earth surface is the *epicenter*. The central angle between this radius and the direction towards the seismic station recorded the arrival of this wave is called the epicentric distance (Δ), and the dependence of the wave transit time (T) on the epicentric distance is the *godograf*. As Earth, as a first approximation, possesses spherical symmetry, then data on the transit time of seismic waves from many earthquakes obtained at numerous seismic stations located worldwide are used to draw the godograf. Its separate points correspond to different seismic rays. Therefore, the seismic waves in elastic media propagate according to the laws of geometrical optics (including the Snellius law of refraction). In this regard it is possible to show that the ratio holds true along any ray:

$$\frac{r \cdot \sin i}{\upsilon} = C = const,$$

where $\boldsymbol{i}$ is the angle between the ray and the $\boldsymbol{r}$ radius at this point; $\boldsymbol{\upsilon}$ the seismic wave speed; $\boldsymbol{C}$ the parameter of the ray which remains for all the points of the same ray.

The parameter of the ray is found from the godograf, since $C = \mathrm{d}T/\mathrm{d}\Delta$; its value is equal to the r/υ ratio at the point where $\sin i = 1$, i.e. at the point of the deepest depth of the ray. The problem of seismology consists just in passing from the godograf (i.e. the Δ dependence of T) to the dependence of the speed of seismic wave propagation on the depth of penetration into the Earth body or on the distance to the center (r), since information of the internal structure of Earth is given by the depth distribution of seismic wave speeds (*the speed cut of Earth*). A network of seismic stations to record the waves coming from various areas of Earth began to be formed in the early 20th century. Primary (P) and secondary (S) seismic waves have appeared to be observed only before $\Delta = 103°$. Between 103 and 142° they disappear and a "shadow zone" is formed. At long epicentric distances (Δ = 142–180°) a later P-wave appears, while the S-wave is absent. To explain these data, an assumption has been made that approximately at half the distance to the Earth center there exists an interface below which P-waves are sharply decelerated and S-waves disappear at all. As is seen from the S-wave speed expression, these oscillations cannot propagate in a liquid media since it does not resist to changes of its shape and its modulus of rigidity is zero. The obtained results indicate the existence of a liquid *core* with a high density at the center of our planet. Above it, the shell or *mantle* of Earth is located, whose substance is solid. The depth of the mantle—core interface (2,900 km) was first estimated by the German seismologist Gutenberg in 1914. On the Gutenberg discontinuity, υ_p jumps from 13.6 to 8.1 km/s, and the S-wave speed falls from 7.3 km/s to zero (Figure 6). Another distinct interface was found by the Yugoslavian seismologist Mohorovičić. At it υ_p and υ_s jump up. The layer located above the Mohorovičić discontinuity is called *the crust*. The thickness of it is various in various parts of the Earth surface. It changes from several kilometers under the ocean bottom to several tens kilometers under the continents. As a result, the Mohorovičić discontinuity is almost a specular reflection of the terrestrial surface relief. The crust differs by not only thickness but also its structure and age.

The top part of the continental-type crust is made of granites and related rocks and called as the "granite layer". A layer with a higher density and elasticity is located underneath. Seismic waves propagate quicker in it. It is conditionally called the "basalt layer", and the interface between these two layers is Conrad's discontinuity. The oceanic type of the crust has a much smaller thickness and consists of the basalt layer only. The geographical areas of the continents and oceans not always coincide with the areas of location of the corresponding-type crust because, owing to tectonic movements, some parts of the terrestrial surface, being now under water, earlier were land, and *vice versa*.

Earth's mantle begins underneath the crust. By the nature of the speed curves of seismic waves it can be divided into three parts, namely: the *top* one (to 400 km, υ_p and υ_s slowly grow or even decrease), the *transition layer* (400–900 km, υ_p and υ_s rise quickly), and the *bottom* one (900–2,900 km, smooth gradual increase in the speeds of seismic waves).

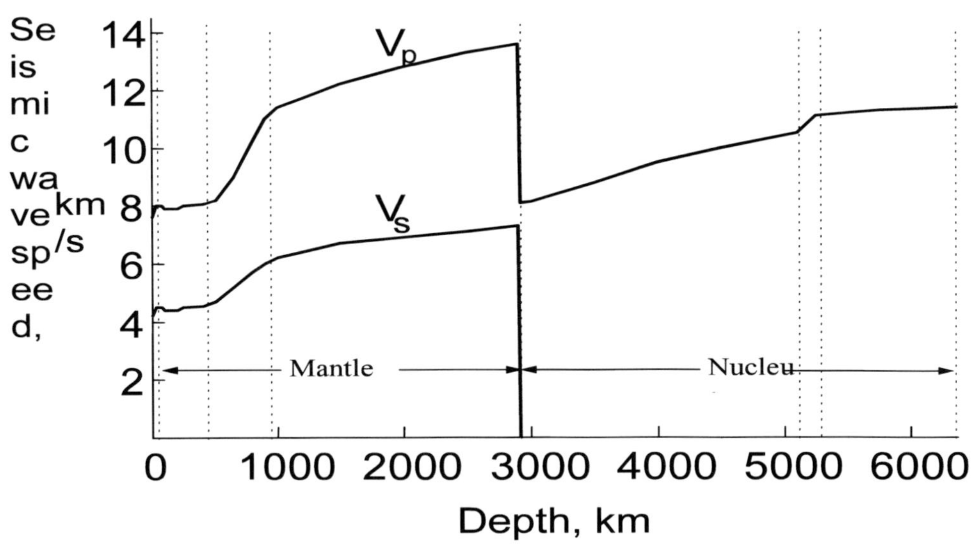

Figure 6. Speed distribution of longitudinal and transverse seismic waves inside Earth.

The mantle plays a very important role in the Earth existence and development. According to modern representations, the crust, directly interacting with live organisms and being part of the biosphere, is a product of substance release from the Earth mantle as a result of bringing the easiest and fusible components up. The centers of large earthquakes are located in the mantle. The top layers of the mantle, as well as the crust, consist of silicates and aluminosilicates differing by structure and crystal lattice shape as depends on conditions (pressure and temperature mainly). The bottom mantle, apparently, except the high-density modifications of silicates and aluminosilicates, contains iron, magnesium, silicon ... oxide mixtures as well. The question of the composition of the inside layers of Earth is solved on the basis of comparison of the physical parameters of each layer with similar characteristics of various rocks obtained in experiments with high pressures and temperatures to model the conditions on various depths. Analysis of the rocks thrown out from the corresponding depths while volcano eruptions, and research of the composition of meteorites (though it is necessary to treat the meteoric analogy with a certain care) give certain information as well. The majority of meteorites originate from the bodies moving in the so-called asteroid belt located between the internal and external planets. Their composition may differ from the Earth one owing to chemical differentiation of substance in the protoplanetary cloud depending on the distance from the Sun.

The part of the globe located deeper than 2,900 km is called as the core which, in turn, is subdivided into three layers, namely: the *outer* one (to a depth of 5,100 km, gradual increase of v_p), the *transitive zone* (the thickness of 100–200 km, a jump of v_p), and the *inner core* (v_p grows slowly) which occupies the most central part of our planet. The high density and high metal conductivity are important properties of the substance therein. It is believed that the core mainly consists of iron, which is molten in the outer core. In the inner core, as appears from seismic data, the substance becomes solid again.

These ideas of the core composition agree with the fact of the existence of iron meteorites. Experimental determination of the iron density under pressures to model the conditions inside the core has allowed assuming the presence of impurity of some easier component (e.g. silicon). According to another hypothesis, the core consists of nearly the same substances as the bottom mantle does, but silicates gain metal properties under high pressure: their density and electric conductivity sharply increase. However, recently this hypothesis began to be called in question since at compression of silicates by a pressure of few million atm no density jump has been detected, though some phase transition proceeds with an increase in conductivity by 5–6 orders of magnitude.

Mechanical Earth Models

Realistic ideas of the density and pressure distribution inside our planet became possible only after the emergence of seismic data to show the existence of several shells with different properties. Henry Cavendish's determination of the gravitational constant value allowed to find the mass of Earth (1798), and its average density at a known volume.

The latter has appeared to significantly exceed the density of rocks (granite and basalt – 2.8 and 3 g/cm^3, respectively) and, therefore, the density of Earth must increase with depth. The *density models* of Earth are underlain by the knowledge of the values of the seismic wave propagation speeds at different depths because the density enters into the expressions for v_p и

υ_S. However, these data are not enough. Leason H. Adams and Erskine D. Williamson contributed most essentially to the design of realistic model of density changes inside Earth. Now the density in the mantle is considered to gradually grow with depth, most quickly in the transition layer. On the border with the core the density jumps almost twice (from 5.5 up to ~10 g/cm^3). In the core the density monotonously increases again, but there is a small jump between the outer and inner cores, speaking for a change in the phase state or the core substance composition on the depth of 5,100–5,200 km. At the core center the density is estimated as 12.5–13.5 g/cm^3. Knowing the sizes of various shells of Earth and the density of their substance, it is possible to estimate the *pressure* inside Earth. It gradually rises, reaching 1.5 and 3–4 million atm on the border with the core and at Earth's center, respectively.

Thermal Model and Earth Radioactivity

Clarification of the *temperature* change as a function of depth is one of the most complex problems of the internal Earth structure. Science still has no techniques of its direct measurement. The thermal mode of the terrestrial surface is determined by two heat flows going from the Sun and from the subsoil of Earth. The former exceeds the latter by few thousand times; however, owing to the low heat conductivity of the terrestrial substance, only a small fraction of the solar energy gets into Earth — at a depth not exceeding few tens meters. The power of the thermal flow released by Earth is estimated as ~10^{21} J/year. A question of the source of this energy thus arises. In the mid-20th century the thermal energy of Earth was believed to be the residue of that heat which Earth had grasped from the Sun at the stage of its formation. Further, universal recognition was gained by the hypothesis of the Earth emergence from a gas-dust cloud possessing a relatively low temperature, and then it was self-warmed up due to the course of some internal processes. These are inelastic collisions of solid particles during the planet formation period, when a fraction of the kinetic energy passed into heat, and the planet's compression under the weight of the newly formed substance. The warming up could occur during the subsequent eras as well, e.g., owing to the still continuing process of differentiation of substance by specific weight. As a result, a planet with the modern density distribution appeared from the originally homogeneous Earth, where the heavy substances lowered to the center and the light ones rose up. However, the greatest contribution to the thermal balance is brought by the energy released while disintegration of the radioactive elements constituting the substance of Earth .

It has been experimentally established that the temperature increases with depth by 15–20° per kilometer in the upper layers of Earth. Estimation of temperature at longer distances from the surface is rather difficult, and the question of the temperature of the terrestrial depths is among those least studied in geophysics. To solve this question, first of all, geologists try to estimate possible temperature limits at various depths. They provide for the fact of passing of seismic S-waves through the whole thickness of the crust and mantle, caused by the temperature of these layers being lower than the melting point of substance (though the existence of volcanoes throwing up molten rocks indicates the existence of such melts in both the crust and the upper layers of the mantle). Such melts can exist as separate centers (chambers with melts). The temperature of the mantle is believed to be ~1,200ºC at a depth of ~100 km. Then it grows with depth, taking on intermediate values between the melting point of the mantle substance under the depth-corresponding pressure (the upper limit) and the temperature caused by its adiabatic compression under the overlying layer weight (the lower limit). If the core is considered to mainly consist of iron (being liquid in the outer core and solid in the inner core), then the temperature in the outer and inner core must be higher and lower than the iron melting point, respectively. As a result, the temperature in the core changes slightly, which agrees with the ideas of its metal nature and the resulting high heat conductivity. At the Earth center the temperature reaches ~5,000°C.

Radioactive elements are components of some rocks and the internal shells of Earth as well. Their concentration is rather low. It differs in various geospheres. The radioactivity of the deep layers is judged by studying volcanism products, based on hypothetical calculations of the temperature mode inside Earth, and by research results of the structure of meteorites of various types (stone, iron-stony, and iron ones) as well. The content of radioactive elements in the crust is higher than in the mantle and core. Heavy radioactive isotopes have large volumes and are thus pushed out from the densely-packed crystal lattices of silicates. In the course of the Earth substance differentiation they, together with the light fraction, floated up, while the dense refractory (ferruterous) substance depleted with radioactive elements fell down. The distribution of radioactive elements over the Earth geospheres changed at different stages of its history. Now, the process of gravitational substance differentiation in the subsoil proceeds, and the concentration of radioactive elements in the crust and mantle increases and decreases, respectively. This changes the radiating Earth background, which influences the life of organisms inhabiting it. The set of the factors influencing the process of continuous species change of live beings on Earth includes the change in the radiating background bringing to mutagenesis, i.e. to the emergence of hereditary changes in organisms as a result of damage and reorganizations in chromosomes and genes. During the last ~600 million years (since the beginning of the population of the continents) it is possible to resolve 20 mutagenic eras repeating with a periodicity of 30 million years. Their emergence is associated with intense riftogenesis, getting magmatic melts from the subsoil of Earth, enriched with toxic (nickel, vanadium, copper, etc.) and radioactive elements. Under the influence of these substances there occurred jumps in the development of the organic world, one species disappeared and others appeared. The violation of the uniform course of the evolutionary process also occurred under the influence of space factors as well as terrestrial ones. In particular, the active riftogenesis, which resulted in the formation of magmatic melts enriched with radioactive elements, is assumed to be a consequence of asteroid body falling onto Earth. The increase in the frequency of meteorite falling (judging by the crater formation periodicity on the Moon) coincided (by time) with the mutagenic eras on Earth and occurred with a period of ~30 million years as well. Besides, falling of large meteorites could cause global climatic changes similar to those expected in the case of mass usage of atomic weapons ("nuclear winter").

Age of Earth

Radioactive isotopes contained in the crystal lattice of many minerals and rocks (e.g., uranium is component of about two hundred minerals) allow to solve the question of their age, and, finally, of age of Earth. *Radioactivity* accompanied by emission of α- or β-particles and γ-radiation is an intranuclear process. It is not affected by the type of a chemical compound (whose composition contains radioactive element atoms), the state of aggregation of matter, pressure, temperature, electric and magnetic fields, i.e. everything what could change the state of the electronic shells, and, therefore, influence chemical reactions. The nuclei quantity of a radioactive element decreases in time under the inverse exponential law:

$$N = N_o e^{-kt},$$

where k is a constant to characterizes the rate of decay of this radioactive element; N_0 the initial isotope concentration.

To estimate the age of rocks or minerals, decay of various elements is studied, and, first of all, of the uranium $_{92}U^{238}$ and $_{92}U^{235}$ isotopes whose concentration decreases twice (a half-life period) for $4.5 \cdot 10^9$ и $7.1 \cdot 10^8$ years, respectively. However, the products of their decay have appeared radioactive themselves, they participate in the subsequent decay reactions with various rates down to the formation of stable lead isotopes $_{82}Pb^{206}$ (from $_{92}U^{238}$) and $_{82}Pb^{207}$ (from $_{92}U^{235}$). E.g., in the case of U^{238} decay, the formation of a lead isotope is preceded by eight reactions of α-decay and six reactions of β-decay. But the slowest stage, limiting the reaction rate in the whole, is transformation of uranium in the reaction of its α-decay. The techniques of *radio chronology* can be applied if the studied object is a closed system, i.e. no initial radioactive isotopes and final products of their decay have migrated from it for the long geological history.

Besides, there must not be atoms of irradiogenic origin among daughter isotopes. In the presence of such impurities it is necessary to develop special methods of their estimation and to introduce corrections. The initial concentration of the radioactive isotope, as a rule, is unknown. Therefore, age estimation is usually based on calculation of the quantity of the remaining initial (parent) radioactive isotopes N and the formed (daughter) products of their decay D which under the above conditions are related by the ratio:

$$D = N_o - N,$$

from whence, taking into account the equation of radioactive decay, it follows that

$$t = \frac{1}{k}\ln\left(\frac{D}{N}+1\right).$$

The age of Earth is obviously longer than the age of the most ancient rocks. In the history of Earth there were two stages, namely, the pre-geological and geological ones. The pre-geological stage begins since the origin of protoplanets and finishes with the emergence of the crust, the formation of the water and air shells of Earth. In those days, much more heat than now was released in the subsoil of Earth, the crust was not isolated yet, and the radioactive elements located more deeply (on the average). The magmatic activity was high, and the formed rocks melted again. The geological history began ~4.5 billion years ago, which corresponds to the age of the most ancient rocks found anywhere on Earth. For a long time it was considered that they had an age of ~3.8 billion years (the Okhotsky massif, the Western Greenland, and Antarctica). However, an even more ancient rock (4 billion years) was found in Australia in 1983, and then – a rock of an age of 4.2 billion years was discovered.

Estimation of the Earth age is also possibly according to data obtained at analysis of meteorites. They are supposed to be formed from the very gas-dust cloud as Earth was, and at the same time. The primary differentiation of the substance on Earth and the substances of meteorites was identical since the formation of the whole planetary system occurred at the same time and under common laws. The conducted measurements in iron meteorites have allowed determining the isotope composition of primary lead. Decay of the uranium containing in stone meteorites leads to the emergence of radiogenic lead. By the radiogenic-to-primary lead ratio, conclusions about the age of Earth are made.

Geomagnetic Field

The hypothesis of the existence of a metallized Earth core explains the origin of the magnetic field of Earth. The modern theories of this geomagnetic field are based on the «dynamo effect» hypothesis put forward by Frenkel in 1947. In the liquid metal core, under the influence of heat released at decay of the radioactive elements contained (e.g., K^{40}) or owing to some other processes (e.g., gravitational differentiation leading to the immersion of iron and the emersion of silicon), convective flows of substance appear. These flows are conductors moving in weak magnetic fields. In these conductors, induction currents appear to induce their own fields strengthening the process, i.e. a mechanism occurs similar to the dynamo machine operating by self-excitation. Various convective cells induce magnetic fields of opposite sign. By summarizing, they, as a result, produce not so strong field of Earth. Magnetic fields are widespread in the Universe. They have been detected in the Sun, stars, plasma clouds moving in space, the majority of planets of the Solar system.

The *geomagnetic field*, as a first approximation, can be considered as the field of a permanent magnet placed almost at the center of Earth (the magnetic center of Earth is displaced from its mass center by 436 km towards the Pacific Ocean), whose axis makes an angle of 11.5° with the Earth rotation axis. This causes the discrepancy between the geographical and magnetic poles (the magnetic inducement discovered by Columbus in 1492). Except the main field whose sources are located inside the Earth core, there are additionally an abnormal magnetic field induced by magnetized rocks located in the crust and an external field whose sources are located in the circumterraneous space. Their contribution to the field observed on the terrestrial surface is 95.4 and 1%, respectively. The magnetic field of Earth plays a very important role in the life of our planet, by protecting its surface from high-energy charged particle beams going from the Sun («solar wind») and from space. It is also a unique reference for sea, air and satellite navigation. The whole area of the geomagnetic field action (the circumterraneous space filled with charged particles moving in Earth's magnetic field) is called *the magnetosphere*. Under the influence of solar wind, the magnetosphere accepts a drop-like shape: it is compressed from the side turned to the Sun and stretches to distances of 60–90 thousand km from Earth; and it is strongly extended on the opposite side to form a magnetic tail of few million kilometers. When the Sun activity increases and the solar wind strengthens, the magnetosphere is condensed, the magnetic field of Earth undergoes short-term distortions («*magnetic storms*»). Charged particles, getting into the magnetosphere, are concentrated in its separate areas called *radiating belts*. There is the external electronic belt distant from the planet by 20–40 thousand km, and the internal proton one, closer to Earth (3–5 thousand km).

Studies have shown that the geomagnetic field varies in time and space. E.g., the magnetic pole moves by 20.5 m per day, or by 7.5 km per year in the Northern hemisphere. Now, it locates in the north of Canada, and in 2185 it will coincide with the North geographical Pole and the inducement will vanish. Changes in the location of the poles are explained, on the one hand, by changes of the field itself, and, on the other hand, by the continental drift, moving of the crust and the top layers of the mantle relative to the internal parts of Earth. Such motion appears possible due to the existence of a layer of low-speed seismic waves, *the asthenosphere*, in the top mantle at depths from 50–80 to 250–400 km, where the substance is amorphous at a temperature close to its melting point, possesses plasticity and a lowered viscosity. There is the lithosphere above the asthenosphere, which is a solid external shell of Earth, including the crust and the top part of the mantle.

Changes in the geomagnetic field can be judged on the basis of studies of the magnetic properties of those rocks which possess "magnetic memory" of the ancient geomagnetic field when they were formed. The property of the rocks containing ferromagnetic materials to be magnetized in their formation under the influence of the Earth magnetic field and to keep the acquired magnetization when cooling down to a temperature below the Curie point (≈750°C for iron) is called as *paleomagnetism*. Studies have shown that not only the location of the magnetic poles changed, but the sign of the magnetic field was repeatedly switching in the past. It is indicated by the opposite magnetization of different-aged rocks. The magnetic screen protecting Earth from charged particle beams disappears in the period of field *inversion*. Space radiation reaching the terrestrial surface influences live organisms and causes their mutation, which entails the spasmodic nature of the organic world development on our planet. Polarity switches of the geomagnetic field occurred through time intervals from several millennia to millions years. Above one thousand inversions have been counted for the last 600 million years. They are believed to have played an important role in the formation of the man when, owing to the magnetic field inversion 2.8 million years ago, Homo habilis replaceed Australopithecus. Then, due to new inversions, the replacement row continued: Homo erectus, Pithecanthropus (about 500 thousand years ago), the Neanderthal man (200–60 thousand years) and, finally, the Kromanian man (~60 thousand years), our immediate ancestor.

8.2. Geodynamic Concepts

The crust formation mechanism is a debatable problem concerning the internal structure of Earth. There is still no enough information to design a sound theory of the development of Earth and its crust but some hypotheses exist trying to solve these problems.

During the development history of geological and geophysical ideas, several geodynamic concepts (ideas of the dynamics of Earth as a complex system evolving over time) have appeared.

The concept of the expanding (growing) Earth. Historically, this concept being one of the very first hypotheses (Lomonosov, Hutton, 18th century AD) well explains such processes and phenomena in the Earth history as the origin of oceans and crust cracking. According to some data, for the post-Carboniferous period, the diameter of Earth and the surface area could have increased more than by a third and more than twice, respectively. No causes of this phenomenon have been revealed as yet. Besides, changes in the key geophysical parameters of our planet require an assumption of changes of other orbital parameters of the celestial body with all logically following consequences.

The contraction concept (from the Latin *contractio* – tightening, compression) is opposite to the previous one by sense. According to this concept (Elie de Beaumont, Kant, Laplace), our planet, originally being in a molten state, was gradually cooling down. Its volume decreased and the crust wrinkled up (to be often compared to the peel of a drying apple), more in more plastic areas containing the layers of sediments, i.e. in the geosynclinalic belts compressed by more rigid blocks of the adjacent craton platforms. Such a mechanism explains the cooling of the subsoil, the crust lowering, the decrease in the volcanic activity intensity, the continuous growth of the sedimentary cover because of the prevalence of immersions over raisings, and the progressive growth of hollows. These representations meet certain objections as well. First, according to much data, Earth originally was not molten but cold and was warmed up later, owing to radiogenic heat accumulation (the maximum of warming up is supposed not to be reached yet). Second, the account of magmatic processes is insufficient; they do not reduce to compression of the subcrustal substance when cooling only. Third, the periodicity of tectonic activity, the brevity of the activation eras, and the duration of the stable eras in the geological history of Earth are not explained. The selectivity of tectonic movements (the analogy with the drying apple peel assumes the uniform folding of the surface) is also unclear. At last, there are calculation data (Magnitsky et al.) of the cooling rate, less than a degree for one million years. With such a small value, the process of cooling cannot essentially affect the excitement of tension in the crust. Therefore, the contraction hypothesis in its classical form could hardly be recognized as well-founded though the Earth compression idea should be considered in the solution of geodynamic and geophysical tasks.

The pulsation concept (from the Latin *pulsatio* – pulsation, palpation) can be considered as a certain compromise between the expansion and contraction hypotheses, it is based on the idea of the compression and expansion phase alternation in the history of Earth (Rothpletz, Grabau, Bucher, Umbgrove, Obruchev, Usov, the first third of the 20th century). At the stages of expansion, radial (vertical) movements prevail. As a result, the crust stretches in the mobile belts to form geosynclinals; the crust protrudes in stable sites, being broken by horsts and grabens. The expansion phases cause magmatizm activization as well. The pressure upon the magmatic layer is supposed to decrease with increasing of expansion, the magma gets thinner, the formation of intrusions and volcanic activity are intensified. At compression, tangential movements prevail to entail crumpling of geosynclinalic deposits into folds and a clumpy folding of the platform base, overthrusts are formed, the regional metamorphism amplifies. The causes of compression are heat release at the previous stage of expansion owing to its carrying out by lavas, gases, vapors etc., and gravitational consolidation of the internal areas of Earth. The expansion causes are possibly related to the magma getting thinner. Now, the pulsation hypothesis is actively developed by Milanovsky who has synthetically combined the compression–expansions processes, riftic zone and volcanism activation, and the stage of transgression–regressive processes.

The concept of continental drift. The idea of possible horizontal moving of the continents appeared owing to the appreciable associativity of the contours and geological features of South America and Africa, North America and Europe, South America and Antarctica. The hypothesis was formulated by Taylor (1910) and Wegener (1915, in his book "The Origin of Continents and Oceans"). It is supposed that the modern continents somewhere (more than

200 million years ago) formed a uniform supercontinent Pangaea. Such a hypothesis proves to be true by the surprising similarity of the outlines of the western and eastern coastlines of the Atlantic Ocean, the uniformity of the geological structure of some adjacent continents (e.g., the unity of the granite massifs of Brazil and Africa), the common nature of the ancient fauna and flora. The icing occurred on Earth 300 million years ago covered so various areas that any explanation of this phenomenon would meet great difficulties. The same can be told about the location of the Earth magnetic poles during the last eras as well. According to paleomagnetic data obtained on several continents, they do not coincide with each other. But everything is arranged if the continents are brought together. Nevertheless, the popularity of this concept has decreased because, first, the geophysical mechanism of such moving of the continents has not been established, and, second, deep breaks going to the mantle and keeping their location for quite long time have been discovered.

New global tectonics (tectonics of plates, plate tectonics, neomobilism). In the 1950–60s, the ideas of horizontal movements revived on a new qualitative basis after the discovery of underwater ridges stretched along the axial zones of the oceans (median oceanic ridges). By the crests of these ridges, located are deep stretching cracks (riftic zones) which the mantle substance arrives onto the terrestrial surface through. The rifts are connected with each other into a uniform all-planetary system. The age of the oceanic bottom has appeared rather short (150–160 million years) and it increases while moving off from the median oceanic ridges whereas the average age of the continents themselves usually exceeds 2–3 billion years. The newly arriving mantle substance pushes aside the earlier released one. As a result, *spreading* of the ocean bottom towards the ridges occurs. According to the spreading hypothesis, the expansion of the oceanic bottom is caused by convective flows of substance in the top mantle whose hot substance is taken out outside at the locations of the rifts. For 200–300 million years, the whole spreading oceanic bottom is renewed, and the former rocks plunge into the mantle in the so-called subduction zones (ocean trenches located near island arches). In these areas, the oceanic lithosphere immerses by first moving under the island arch and then moving farther, under the continent. This process probably causes the continental drift. For example, North America and Europe gradually move apart from each other as the new oceanic crust is formed between them. The main motive power in plate tectonics is the force of draft of the heavy lithospheric block plunging into the mantle. As a result of the continental crust compression by the meeting plates, young folded mountains appear. Mountain formation is explained by that the continental crust has a low density and, consequently, cannot fall into the mantle at plate collisions. At lateral moving of next plates, transverse shifts (transform breaks) occur. Where the blocks of the oceanic and continental crust influence on each other, seismically active continental peripheries appear. The versions of plate interaction are various: collision of two continental plates accompanied by the formation of mountain systems (the Himalayas), sliding relative to each other (the San Andreas break in California), etc. The revealed changes in the location of the continents by paleomagnetic techniques and the geophysical confirmation of the existence of a layer of low-dense substance in the top mantle (the asthenosphere) have served as confirmations of the new global tectonics. At the same time, this concept cannot explain the lack of subduction zones in some areas (e.g., round the Antarctic and African–Arabian lithospheric plates), the absence of a low-speed zone (asthenosphere) under the Archaean boards, etc. The development of the Precambrian geology also shows that this concept is inapplicable at early stages of the crust evolution.

Concept of subcrustal convective currents. In 1924 J. Jolie stated a hypothesis of the influence of the radioactive element decay heat on the geodynamic processes, namely: the heat, accumulating, warms up the bases of the continents, causes basalt melting, and the continents immerse into the molten basalt, which determines activation of transgressions. Then, basalt melts under the ocean, and the continental massifs "float" on it westwards, cool down and rise, causing folding. The continents and oceans thus as though interchange their locations. In 1929 A. Holmes assumed that the non-uniform distribution of the subcrustal radiogenic heat could initiate a system of convective currents in the mantle. The ascending currents cause rupture of the continents, their crushing, and the formation of new oceans; on the borders of the continents and oceans reverse (descending to the mantle) streams are developed.

The concept of plumes and hot fields. Under plumes are understood narrow columns of the heated mantle substance rising from the deep layers of the mantle. They originate on depths not less than 700 km, their diameter varies from 100 to 240 km, and the speed of rise is ~ 2 m/year. As a result of the plume activity, domes with a diameter up to 1,000 km are formed, whose central parts tower by 1–2 km. Plumes are related to (sometimes with full identification) the notion of hot points, i.e. sites of the terrestrial surface sharply distinguished by strongly increased volcanic activity. Plumes are observed inside plates (of the Hawaiian type) and on the extending borders of plates (of the Icelandic type). They behave as solid plastic (probably, partially molten) bodies and rise similarly to hydrochloric diapers. Their substance is lighter than the surrounding rocks, and the viscosity is low enough (of the order of 10^{19} P) for occurrence of currents. The gradual reduction of the pressure in the plume substance results in an increase in the melt fraction, which promotes further rise and causes volcanic activity in the field of hot points (the lighter molten part is separated). Plumes are supposed to be part of a complex global system of convection flows in the liquid outer core and mantle. Two models of thermal gravitational convection are possible, namely, over the whole thickness of the mantle from the lithosphere (30–100 km) up to the core–mantle interface (2,900 km); and in two layers (in both the top and bottom mantle). Mantle plumes may arise at three levels, namely: in the top mantle (in particular, at melting of the subducted fragments of the plate), in the transitive layer of the mantle, on the mantle–core interface.

The concept of fixism. The concept of fixism assumes invariance of the mutual arrangement of the continents in the Earth history. The oceans were formed on the place of continents in the Mesozoic-Cenozoic stage of the Earth development. For a long time, Earth was warmed up by internal radioactive heat sources on depths of hundreds kilometers. As a result of partial melting of the ultra-basic rocks from the top mantle the constituent basalts were smelted. Further differentiation of these basalts resulted in isolation of an acidic (granite) material therein. So was formed the continental crust covering the whole globe just in the Archean. Only small intracontinental sea pools were developed on the formed crust, with no oceans. Enhancement of warming up resulted in the appearance of centers of complete melting of the ultra-basic mantle substance. The ultra-basic and basic (basalt) melts went upwards along breaks, penetrated into the crust, and were given vent onto the surface to form eclogites. Their density considerably grew, and heavy blocks of the crust, together with the penetrated intrusions, were immersed back into the mantle, melting therein. As a result, a part of the acidic material of the crust was replaced by the ultra-basic mantle material to form the crust of a new type. This process named "basification" has finally led to "oceanization", i.e. formation of an ocean on the place of the continent. The crust renewal process mainly finished in the Cretaceous age, though it proceeds till now here and there, as lowering of the ocean bottom proceeded from their periphery to the axis.

8.3. Basic Stages of Earth Evolution

For billions years of its existence, Earth and its biosphere have passed a complex way of development before getting their modern properties, structure, and composition. Now, geological periodization of the Earth development is elaborated. The pre-geological stage lasted ~ 200 million years (ca. from $4.7{\cdot}10^9$ till $4.5{\cdot}10^9$ years back) is referred to as *Catarchean* (most ancient). The geological Earth history is divided into two unequal stages (Table 6). The most ancient of them, covering about 85% of the time of our planet existence, is referred to as *pre-Cambrian.* It has gotten such a name as being preceded to the first (Cambrian) period of the subsequent megastage called *Phanerozoic* (obvious life) and covers the last 570 million years. Pre-Cambrian is subdivided into two eras, namely, Archean (the initial, ancient era) and Proterozoic (the earlier life) ones. Phanerozoic is divided into three eras: Paleozoic (ancient life), Mesozoic (middle life), and Cenozoic (new life), each of which counts several periods.

At the first, Catarchean stage, Earth was a rather cold and tectonically passive planet. In that epoch it only gradually got warm owing to decay of radioactive elements and dissipation of the tidal energy. The Moon played a significant role in that period. According to some representations, it was formed at destruction of a more massive planet (the proto-Moon) captured from a close orbit just during the Earth growth.

Table 6. Main Stages of Earth Evolution

Eon	Era	Period	Beginning (mln years ago)
Catarchean			4,700
Pre-Cambrian	Archaean		4,500
	Proterozoic	Lower	2,600
		Upper	1,650
Phanerozoic	Paleozoic	Cambrian	570
		Ordovician	500
		Silurian	440
		Devonian	410
		Carboniferous	355
		Permian	285
	Mesozoic	Triassic	230
		Jurassic	195
		Cretaceous	140
	Cenozoic	Paleogenic	65
		Neogene	25
		Quaternary (Anthropogenic)	1.5

The most part of the substance of that planet dropped onto Earth. Degassing of Earth in the Catarchean almost did not occur. Therefore, the terrestrial atmosphere was extremely rare and mainly consisted of inert gases, since all chemically active gases were sorbed by its porous ground and gradually buried in the terrestrial bowels. The hydrosphere was absent in that period.

Release of water vapor and other atmospheric gases began only 600 million years after the Earth formation. In the Archean it already was enveloped with a dense atmosphere consisting of endogenic gases (carbon oxides, ammonia, methane, hydrogen, anf water vapor) causing a rather strong hotbed effect. In the same epoch, first shallow isolated sea pools appeared, but already by the end of Archean, a uniform but still shallow global ocean was formed. This promoted carbonic gas fixing as carbonates. As a result, at the very beginning of the early Proterozoic (about 2.5 billion years ago) the CO_2 content considerably decreased and the average terrestrial temperature decreased.

The chemical composition of the atmosphere practically has not changed during the last several millions years. This can be explained by its regulation by biological processes occurring towards optimization of the biosphere development conditions.

8.4. Global Climatic Concepts

The Earth climate depends on both space and planetary factors. The concepts now known are united into two groups: *astronomic-physical* and *geologic-geographical* ones.

The astronomic-physical concepts associate climate variations with oscillations of the quantity and composition of solar radiation. They are divided into two subgroups. The first one (*Milanković's paradigm*) starts with the principle of solar radiation constancy and the variability of the terrestrial surface location relative to the solar rays. The second subgroup is associated with changes in the Sun radiating ability.

Milanković's paradigm. The ideas of non-uniform distribution of the solar radiation over the terrestrial surface owing to changes of the terrestrial orbit elements were first offered by Henry Kroll in 1875. He hypothesized the relation of the periodic changes of the location and volume of the polar ice caps caused by climate fluctuations with the variations of the terrestrial orbit's eccentricity and the inclination of our planet's ecliptics. Variations of the Earth rotation axis and the terrestrial orbit elements influence the seasonal and latitude distribution of solar radiation over the terrestrial surface. Changes in the eccentricity of Earth's orbit promote changes in the solar energy quantity. Distortions of the two other elements lead to the latitude redistribution of solar radiation. Changes in the insolation determining the thermal condition of the terrestrial surface cause the intensity of evaporation and condensation of atmospheric moisture, the force and direction of winds and sea streams, the winter and summer duration, the location of the upwelling zone. Therefore, fluctuations of the warm and cold, dry and damp types of climate take place. Recently, a hypothesis of the existence of "Milanković's cycles" for the Sun, i.e. cycles of precession (95 million years), ecliptics (190 million years), and eccentricity (570 million years) of the orbital parameters of the Sun, has been offered.

Öpik's hypothesis is based on an assumption that at the development of a star similar to the Sun, the hydrogen to helium transition creates instability and mixing zones therein, which leads to temporary weakening of the solar radiation (to a cold snap). After balance restoration, the intensity of solar energy increases (warming). The main demerit of this hypothesis is the absence of any observations and theoretical calculations to verify the fact itself and the periodicity of the Sun's "blinking". The solar activity periodicity is estimated as 70 or 200 years by different authors, which finds reflection in the climatic fluctuations fixed in the annual rings of modern and fossil trees.

Predtechensky's hypothesis. In 1950 it was hypothesized that the solar activity influences the basic circulation mechanisms in the atmosphere through condensation processes: west–east transfer and meridional exchange. Under strengthening of the solar activity, the circulation in the atmosphere increases. The equatorial and tropical zone expands, but the average temperature therein decreases owing to the increasing cloudiness. The Polar Regions expand owing to advection of the warm air masses of the lower latitudes. The temperate climate zones are shrunk. The interzonal exchange of air masses weakens, and the temperature equator—pole gradient decreases, especially during the wintertime. The climate in all the circulating belts becomes less continental and monotonous over vast spaces. The extreme types of climates, the desert and Arctic ones, disappear. When weakening of the solar activity, stationary-type processes become prevailing. The mid-latitude zones reach their maximum development, and the other zones reduce. The meridional exchange weakens, and west—east transfer becomes absolutely prevailing. The continentality of climate increases, and the temperature equator—pole gradient reaches its maximum value. The contrast between climatic zones becomes sharp. P.P. Predtechensky classifies the glacial eras as transitional states when meridional invasions and west—east transfer appear more or less equivalent and when temperature decreases, especially in the summer, and the precipitation is abundant. Such conditions favor to snow accumulation and establishment of a glacial era. The plurality of glacial eras is explained by superposition of solar activity cycles of various duration and amplitude.

The geologic-geographical concepts can be also divided into two subgroups, namely: few ones explain climate variations by changes in the composition and properties of the atmosphere; others by changes in the terrestrial surface.

The hypothesis of the volcanism role. It should be noted that the considerable fall of solar radiation (by 10–20%) was a consequence of the catastrophic eruptions of volcanoes: Krakatau (1883) and Mount Katmai (1912) (18 and 21 km^3 of friable volcanic products thrown out, respectively). Modern observations show that great eruptions of volcanoes are followed by cold years. However, this periodicity of volcano eruptions and the subsequent cold snaps contradicts the known astronomical cycles, which does not allow to surely estimate the volcanism role as a climate factor in planetary scale. It is only possible to reliably judge on a local impact on the climate.

The hypothesis of the terrestrial relief role. This hypothesis is based on the interrelation of the climate and Earth relief. First, the relation of the areas of the continents and the ocean is reflected in the general type of climate and circulation in the atmosphere. Second, of certain importance is the geographical arrangement of the main continental blocks. If they are concentrated in high latitudes, the climate will be considerably continental, than in the case of their placement in low latitudes. Third, the average hypsometric level of land has certain climatic significance. Its increase entails atmospheric pressure reduction, increased evaporation, and a common temperature fall. At falling of the average hypsometric level the climate of the continents becomes warmer and damp. Fourth, the role of high ridges-climatesheds blocking atmospheric circulation is considerable. The Himalayas dividing the monsoonal area of Hindustan and the dry area of Tibet can serve as a modern example. There are considerations close in sense of the relation between the icings and the orogenic periods, when the land-occupied area sharply increased with the mountain height. However, the history of geological development and the present give many examples disproving any direct link between paleogeographic and geographical, paleoclimatic and climatic changes. Icing accompanied the Variscan and Alpine phases of mountain formation (orogeny) only. There is no evidence of icing for all other orogenic phases. Sharp alternation of glacials and interglacials in the Quaternary Period occurred with an invariable relief.

The hypothesis about the World Ocean role. The ocean surface receives heat owing to solar radiation absorption and gives it by effective radiation (42%) and evaporation (51%). Therefore, the World Ocean, being the main source of moisture and a thermal energy keeper, changes its temperature slightly and slowly.

The hypothesis about the role of raisings of the oceanic bottom. The formation of underwater ridges leads to disturbance of the water mass circulation. This entails isolation and fall of temperature in the polar pools. This hypothesis offered by Rukhin in 1958 is exemplified by the linkage between the Quaternary icing with the raising of Thompson's underwater ridge (the Atlantic Ocean). The risen ridge blocked the Gulf Stream access to the Northern Polar pool which began to cool and to be covered with ice. The cold masses of dry air formed over this pool moved towards the equator and, adjoining to the Atlantic warm air masses, promoted the formation of precipitation (mainly snow). These factors caused icing development in Europe and North America. At the subsequent lowering of Thompson's ridge, the warm waters of Gulf Stream got to the north, warming up the Northern polar pool. The warming up of the Northern pool was accompanied by degradation of the snow and ice covers.

8.5. Carbon Dioxide, Hotbed Effect, and Earth Climate

There is a close relation between the atmosphere pollution and climatic factors which are determined by the thermal balance of our planet in many respects. The slightest disturbance can lead to opposite ecological disasters — to the world icing or to not less dangerous sharp warming. Changes in the intensity of solar radiation, in the location of the terrestrial surface relative to the Sun, and in the atmosphere composition could be causes of thermal balance disturbance. About 99% of the solar energy available to Earth is concentrated within the wavelength range from 170 to 4,000 nm, and the majority of atmospheric gases are transparent for solar radiation, having almost zero absorption factors within this range. However, in the mid-19th century, the English physicist J. Tindal established water vapor and carbon dioxide to absorb IR radiation. As a result, the atmosphere appears to be almost transparent for coming solar radiation but not transparent for the long-wave IR radiation from the terrestrial surface. The said gases absorb IR radiation and thus heat up. Due to this effect the air temperature near the terrestrial surface is higher than it would be in the absence of these gases. The average air temperature near the terrestrial surface now is about +14°C. Calculations show that it would be close to -20°C if carbon dioxide disappeared from the atmosphere. Carbon dioxide and water vapor create the so-called "hotbed effect", i.e. play the same role as the hotbed glass: absorbing the heat radiated by Earth, they increase the temperature of the lower troposphere. Some other components of the atmosphere of both artificial and natural origin also belong to the hotbed gases, namely: methane, freones, ozone, nitrogen oxides. The total number of the hotbed components in the atmosphere, whose contents is mainly determined by anthropogenous activity, is estimated as few tens. The hotbed effect of each gas depends on the efficiency of IR radiation absorption by its molecules, concentration, and lifetime in the atmosphere. Water vapor contributes mainly to the hotbed effect. However, it (as well as any other substance) absorbs not the whole range of IR radiation but has so-called "transparency windows" within 3,000–5,000 and 8,000–12,000 nm. Carbon dioxide and some other hotbed gases supplement the effect of water vapor, absorbing IR radiation at the wavelengths corresponding to its transparency windows. The main role in this process belongs to CO_2.

Many researchers have tried to estimate what climate changes the atmosphere enrichment with carbon dioxide would lead to. In 1896 Svante Arrhenius calculated that an increase in the CO_2 concentration by 2.5–3 times would raise the average air temperature near the terrestrial surface by 8–9°C. Other estimates were made as well. Other researchers also consider that at an increase in the carbon dioxide concentration, temperature would inevitably increase. However, high distinctions in the quantitative estimates of this increase (from 0.5 to 5.5°C on doubling the CO_2 content) are observed. As a rule, such estimates are made by modeling of the processes proceeding in the climatic system. These models are usually too approximate and could give incorrect forecasts. E.g., according to many experts, the average annual temperature by the end of the 20th century should have increased by 1–2°C in comparison with the 1970s. However, the actual observed warming for this period was 0.35°C.

The atmospheric carbon dioxide participates in some circulations associated with the activity of green plants and the photosynthesis reactions proceeding therein, and also with gas exchange processes in the lithosphere, hydrosphere, and atmosphere. During many millions years, volcanic eruptions were the main supplier of carbon dioxide into the atmosphere. In the geological past, much many volcanoes functioned on Earth than now. A close relation between the volcanic activity level and the carbon amount in the sedimentary rocks has been established. During the era of volcanism strengthening, the CO_2 concentration in the atmosphere increased and this promoted a weight increase of the carbonate rocks formed. The carbon dioxide concentration in the atmosphere during various periods of our planet's history was subjected to considerable oscillations. This is quite possibly the very factor to determine the temperature mode on Earth at different stages of its development. During the history of Earth, periods with the tropical climate and periods of icing were observed, alternating with certain intervals. The maximum of the last icing was 20 thousand years ago only. Then, the icing border did not a little reach that place which now Moscow is on. Then, warming came. Its maximum known as the "Holocene optimum" was 7–8 thousand years ago. Since then, the Earth climate started to become more and more cold and dry. An especially strong cold snap was in the 14th century AD.

Ice contains much less CO_2 than water. Therefore, during icing, the oceans, whose water volume decreases, gradually release the carbon dioxide dissolved therein into the atmosphere. The hotbed effect thus amplifies. The general heating of our planet occurs and the ice age comes to an end. Thawing of ices and sea level rise accompanied by partial absorption of carbon dioxide from the air lead to a new icing, etc. Vegetation contributes to the maintenance of the cyclic nature of these processes. It strongly spreads out in the conditions of a raised content of carbon dioxide, transforms it to various organic and mineral compounds of fossil carbon, and thus lowers the content of CO_2 in the air to reduce the hotbed effect.

The specified cyclic processes proceed against the general attenuation of volcanic activity on Earth, which leads to reduction of the carbon dioxide intake into the atmosphere. As a result, its earlier accumulated stocks were gradually exhausted owing to dissolution in the ocean and absorption by green plants. The CO_2 content in the atmosphere gradually decreased and in the mid-20th century it became lower than 0.03%, which never occurred for the whole geological history of Earth. If the modern weakening of the Earth degassing has not yet finished and the carbon dioxide concentration will continue falling, then, how is considered by Budyko and Yanshin, next icing is probable.

Climate is the average long-term mode of weather. It can be therefore characterized only on the basis of continuous observation during tens years. It has appeared that for the second half of the 20th century the CO_2 content in the atmosphere increased by one third of its initial level (0.028% in 1956 – 0.037 % in 2000) and the average temperature of the terrestrial air layer increased by 0.5°C for the same time. This is much more than for any other previous period. Many people think that now the main cause for the carbon dioxide intake in the atmosphere is burning of fossil fuel to obtain energy. The total carbon dioxide thus formed becomes comparable with what volcanoes (nowadays inactive) earlier gave. Nevertheless, the CO_2 concentration in the atmosphere grows more slowly than could be expected. The matter is that, along with photosynthesizing land organisms, a huge role is played by the ocean in whose waters more (almost hundred times) CO_2 is dissolved than contained in the atmosphere. Carbonic acid formed on CO_2 dissolution in water is fixed by calcium ions as insoluble $CaCO_3$ (limestone). The quantity of carbonates is ten thousand times higher than the modern CO_2 content in the atmosphere. Besides, carbonic acid is consumed by photosynthesizing organisms and passes to the biomass of other inhabitants of the ocean through nutrition circuits. Their died-off remains settle on the bottom, turning into fossil coal and oil with time. The quantity of CO_2 buried with these minerals exceeds its atmospheric contents by one thousand times. The ocean is thus a powerful reservoir regulating the gas composition of the atmosphere. This conclusion has formed the basis for an opinion (very widespread somewhen) that the whole man-produced carbon dioxide is dissolved in the waters of the oceans. However, in the 1930s it was shown that the ocean capabilities of regulating the carbon dioxide content in the atmosphere are limited. This is due to rather slow CO_2 diffusion from the top, well mixing layers into deeper ones. Deforestation has become another cause promoting carbon dioxide accumulation in the atmosphere. In the 20th century, about half the rainforests of our planet has been destroyed; 16–17 million hectares of wood which is the main carbon dioxide into oxygen converter while photosynthesis reactions are lost annually. Huge damage is due to forest fires taking the second place after fuel burning as CO_2 suppliers into the atmosphere.

As the change of the carbon dioxide content in the atmosphere is of both anthropogenous and natural character, a question arises with no rather reasoned answer as yet: whether the modern increase in the CO_2 concentration is a consequence of some natural processes on the background of which the anthropogenous influence plays a secondary role mistaken by us as the primary factor. Besides, mankind still has no objective evidence of Earth being heated due to the increase in the hotbed gas concentration in the atmosphere. To solve these questions, a comprehensive complex consideration of the whole variety of biogeochemical processes in the biosphere in their interrelation and interconditionality is necessary. E.g., it was shown that the average temperature of the terrestrial air layer rose in the first third of the 20th century, then remained at a constant level (or even decreased a little) from 1940 to 1970, and started to raise slowly in the 1980s again. So, in the mid-20th century, despite the increase in the carbon dioxide concentration in the atmosphere and the existence of preconditions for amplification of the hotbed effect, temperature did not change. Studies have revealed natural temperature oscillations with a period of about seventy years. Together with the warming owing to the hotbed effect amplification, these oscillations have led to temperature equalization in the middle of the last century. Temperature oscillations with a seventy-year period are observed in the Northern hemisphere only and most distinctly in the Northern Atlantic. These data agree with the results of oceanographic research. They allow hypothesizing that the Earth climate mainly depends on the ocean stream in the northern Atlantic Ocean. Depending on acceleration or retardation of this stream, the heat quantity transferred from the tropics to the Arctic changes. Such changes in the ocean are just believed to cause the seventy-year temperature oscillations. The given example once again shows the multifactoriality and coherence of processes in the biosphere.

Now the question is actively discussed what consequences the increase in the hotbed gas content in the atmosphere could lead to. Fears are expressed that the increase in the average temperature, especially in the polar regions, could cause fast thawing of the ices of Antarctica and Greenland. As water absorbs solar energy by 8 times more intensely than ice, after thawing of the latter, an additional warming caused by the reduction of the reflective ability of the terrestrial surface (albedo) should develop. This will cause sharp lifting of the World Ocean level to flood the coastal cities and lowlands, which will be accompanied by economic and social shocks. Therefore, reduction of the consumption of coal, oil, gasoline, and other types of fuel is called for. The atomic energy wide usage supporters take advantage of this. They emphasize that nuclear power stations release no carbon dioxide and are therefore environmentally safer. However, the modern level of the operational safety of nuclear power stations does not allow making such unequivocal conclusions. A question arises of how substantiated are the fears of thawing of the Antarctica and Greenland ices. The volume of these ices is known rather precisely — 27 million cubic kilometers. If all this ice thawed, the level of the ocean would have risen above 70 meters. However, judging by the data of the boreholes passed the whole thickness of the glacial board of Antarctica, it was formed more than 30 million years ago and has sustained several eras of Earth climate warming since then, much more intense than expected from the hotbed effect. By the experts' estimates, the lifting of the World Ocean level does not now exceed 0.8 mm per year.

Climate warming because of the growth of the CO_2 concentration in the atmosphere will lead to an increase in evaporation from the oceanic surface and, therefore, to an increase in the precipitation amount. However, the warming process will occur non-uniformly, namely: at the poles more intensely than on the equator. This is owing to that the concentration of water vapor in the atmosphere is rather low in high latitudes because of the low temperatures therein (as vapor is frozen out and drops out as precipitation, increasing the fraction of the reflected solar energy). In such conditions, CO_2 becomes the major hotbed gas. Because of the different extent of warming at the equator and poles, the nature of atmospheric circulation, the force and direction of winds, rains, and ocean streams will change. As a result, the frequency of large temperature anomalies and droughts in the mid-latitude continental regions could rise. And the weakening of the warm Gulf Stream will lead even to cooling in Europe. The meteorologists of the whole world now discuss a question of how the additional precipitation caused by warming will be distributed over areas and seasons. The answer to this question can be apparently gotten from analysis of the situation existed on Earth 7–8 thousand years ago (during the Holocene optimum).

Thus, among the climatology experts there is no uniform understanding of the causes of the modern climatic changes and those of the hotbed effect amplification, and the participation of anthropogenous and natural processes therein. Forecasts of the expected consequences are rather inconsistent as well. Besides, considering the anthropogenous pollution of the atmosphere, it is necessary to mean that its separate forms could lead to consequences of an opposite character. E.g., the increase in temperature because of the hotbed effect is resisted by its possible reduction caused by an increase in the dust content in the atmosphere, i.e. its aerosol pollution. As a result, the temperature will rise or lower depending on which influence (that of carbon dioxide or of the mechanical impurities of artificial origin) will prevail. There are also forecasts of that in our century the Earth temperature will be mainly influenced by various impurities. They will not only counterbalance the carbon dioxide influence but also cause an essential fall of the terrestrial surface temperature.

REFERENCES

Gabdullin R.R., I.V. Ilin, A.V. Ivanov. *Evolution of Earth and Life*. Moscow, 2005. [Russ].

Goldfein M.D., A.V. Ivanov, A.N. Malikov. *Concepts of Modern Natural Sciences*. Moscow, 2009. [Russ].

Chapter 9

Modern Concepts of General and Applied Ecology

Nature operates according to its own laws while the man does according to his ideas of law.

August Lösch

9.1. Concept of the Ecosystem

9.1.1. Trophic Chains

The principal feature of a biocenosis is its ability to maintain steady substance circulation in the biosphere on the basis of food relations among different populations. The food relations among organisms in ecosystems are usually represented as nutrition chains or trophic chains (from the Greek *trophos* — food). All known living things are divided into three groups, namely, producents (those who produce), consuments (those who consume), and reducents (those who reduce or decompose). *Producents* are the organisms capable of synthesizing organic substances from inorganic ones. Typical representatives of such autotrophic organisms are green plants and some bacteria capable of *photosynthesis*, i.e. transforming the energy of solar light into that of chemical bonds in organic substances. The energetic sense of photosynthesis consists in light splitting of some hydrogen donor H_2D and transfer of the hydrogen atoms to some acceptor A with release of the free oxidizer D:

$$H_2D + A \xrightarrow{\textit{light}} AH_2 + D.$$

Some organic compounds, hydrogen sulfide or water can play the role of such a hydrogen donor, and carbon dioxide acts as an acceptor. Most widespread is the process involving water and going in plants, where solar energy is entered into a chain of photosynthetic transformations through the green pigment chlorophyll:

$$6\,H_2O + 6\,CO_2 + 673\,kcal \xrightarrow[\textit{chlorophill}]{} (CH_2O)_6 + 6\,O_2.$$

Photosynthesis is a rather complex reaction consisting of many stages: water photolysis, polymerization, etc. The total energy annually reserved in plants is very high (~10^{24} kcal) but the efficiency of photosynthesis is quite low (about one percent). This is due to chlorophyll absorbing not more than 10% of the energy of the whole spectrum of solar radiation (in the light-blue 400–500 nm and red 610–690 nm ranges), i.e. a rather small fraction of the whole energy of the Sun coming to Earth.

Next link in the trophic chain is *consuments*, i.e. such organisms who utilize ready organic substances from plants and animals as food. They include several trophic levels, namely, first-order consuments (vegetable-eating or herbivorous animals), second-order ones (carnivorous), and third-order ones (predators and parasites). These groups of organisms decompose complex food substances down to simpler ones, synthesize proteins, fats, and carbohydrates therefrom, which then are used for construction of the body and as a power source.

The terminal link in the trophic chain is *reducents* who utilize dead organic substances (decomposing the remains of dyed organisms or waste, e.g. foodstuff unsuitable to further use) as food. These are bacteria, microorganisms, and fungi. They return mineral substances into the environment, ready for consumption by plants and the initiation of a new biogenic cycle.

The food relations in a biocenosis are therefore determined by substance circulation and represent energy transfer mechanisms from one type of organisms to another one. Analysis of the relations between the links in a trophic chain shows that at transition from one link to another the energy accumulated in the consumed mass dissipates; hence, shorter trophic chains are more favorable energetically. Unlike substance circulation, no energy circulation occurs just because energy practically does not come back from reducents to producents. Ecosystems, being open systems, always require fresh energy inflow to function. The Sun is the source of new amounts of energy, whose energy is transferred through the links of a trophic chain and converts from one kind into another one. According to the law of energy transfer and transformation, the trophic chain is a way of one-directed flux of a rather small fraction of highly-effective solar energy through living beings into the environment, as poor-quality thermal energy (waste).

9.1.2. Ecology of Populations and Species

Unlike Charles Darwin's theory of origin of species, the modern representations of biochemical evolution consider *the population* as an elementary (primary) hierarchy level in animate nature, and changes in populational properties are the determining process in biochemical evolution. The life on Earth is characterized by continuity and discrete properties, which manifests itself as the presence of separate individuals whose numbers are divided into groups to form a system of species and populations. Various geographical regions of the biosphere are populated with living beings of various species with certain numbers of individuals. Every separate species of organisms always exists in the biosphere not as separate individuals but as their set occupying a certain territory. In this connection a question arises how to define the notion of “species”. The term “species” is generally used to denote any group of similar objects, which this or that name is given to. This term is frequently used for living organisms as well. The biological concept of species means that a species consists of populations and has its internal genetic unity (gene pool). So, all the individuals of a species have a common genetic program (historically established during evolution) and form a so-called *reproductive community*. In this connection, the male and female individuals of the same animal species react to each other as to potential conjugal partners and find each other with the purpose of reproduction. Therefore, on the one hand, *the species is an ecological unit* and interacts with other species inhabiting the same environment, but, on the other hand, *the species is a genetic unit* containing a huge interconnected gene pool. At the same time, an individual is only a small piece of the gene pool, existing during a short period of time. Any species represents a *protected gene pool* which is expressed in the genes from one gene pool forming harmonious combinations, as they are co-adapted by natural selection. However, it is necessary to note that the biological concept of species has a certain sense only in that case when the populations existing in the same place and in the same time are spoken of. The more distant some two populations from each other are in space and in time, the more difficult is to verify their specific status relative to each other. From whence it follows that *the species are groups of crossed natural populations reproductively isolated from such other groups*.

The notion of "population" occupies a special place in biology and ecology since this is such a natural phenomenon which the real existence of any species of living organisms is connected with. In this sense it is possible to consider a population as the elementary (primary) form of species existence. Under the population is usually understood a set of individuals of one species with a common gene pool, which occupy some space with rather homogeneous habitation conditions during many generations. Inside a population this or that degree of free mixing based on casual and equiprobable crossing of all the types of individuals occurs. Every population has its own destiny in that sense that it is capable (at least, theoretically) of developing beyond all bounds in favorable environmental conditions. Sometimes, groups of spatially adjacent populations may form a so-called *geographical population*. Moreover, neighboring populations of one species, as a rule, differ from each other by certain attributes some of which gradually and naturally change. Such evolutionary changes of attributes can be considered as adaptation of populations to local conditions. To denote changes in any attribute, Thomas Henry Huxley offered the term of *cline* in 1942. Clinality is a result of the action of two opposite forces (competition): first, natural selection which aspires to make every population adapted to its local environment, and, second, the presence of genes aspiring to make all the populations of the species identical. Clines are widespread and met in the majority of microorganism, plant and animal species. And this is clear because such climatic factors as temperature, precipitation mode, evaporation, the number of cold and warm days, snowfalls etc. change, as a rule, naturally and directionally. Hence, the presence of clines points out how closely the organisms and their populations are connected to the external environment, depend on it, and react to its changes. At the same time, it is necessary to mean that the natural-territorial complex enveloping a population may have various features caused by the natural-climatic zone, landscape and its separate morphological parts. Therefore, on the one hand, any population is characterized by its isolation from the other populations of the given species. On the other hand, populations as genetic-ecological systems differ in structure, size, ecological and genetic parameters, and the level of links among themselves.

Under *the structure of a population* is meant any functional parts thereof as a whole connected in a certain order. Such parts are the sexual, age, spatial, genetic and ecological structures. All of them can be considered, in a certain sense, as versions of the genetic structure, since any division of a population during its evolution depends, first of all, on genetic attributes. Usually *the sexual structure of a population* is considered as the male to female ratio in various age groups. Therefore, the sexual structure of a population is determined by the primary (at the appearance of zygotes, i.e. cells formed upon fertilization), secondary (in newborns), and tertiary (at sexual maturity onset) sex ratios. Naturally, there is no sexual structure of populations for lower plants and animals. For the majority of species of living beings the sexual structure of a population is an enough universal characteristic. It is therefore possible to consider that the sexual structure of a population is described by the male to female ratio in various age groups, the fraction of sterile individuals (incapable of reproduction), and the fraction of individuals with different sets of chromosomes. Besides, this parameter is a specific index for every population to reflect the genetic, ontogenetic, and biogeocenotic influences upon specific populations.

Every individual not only belongs to a spatial group but also is a member of a temporary group, i.e. a generation, litter, or age group. That is why every population has a certain *age structure*, which reflects such important processes as the intensity of reproduction, death rate, and the rate of generation change. The age structure of a population depends on genetic features of the species, which, in turn, are determined by particular conditions of the existence of separate populations. Usually, the specificity of age structure manifests itself through the ratio of groups of individuals of different absolute ages (so-called age pyramids), through the ratio of different generations, litters and age groups, through the ratio of pre-reproductive, reproductive and post-reproductive periods, and through the character of growth of individuals. In a normal (stable) population, the number of young is always higher than that of more senior individuals. *Age pyramids* frequently appear rather mobile characteristics, which may change even in one population over rather short periods of time. Besides, two more general characteristics of the age structure of a population are highlighted. First, the reproductive age in some species appears to be so a dynamical characteristic that it may vary from population to population; even inside one population there exist individuals sharply differing by this parameter. Second, in some species the age structure becomes complicated due to non-uniform maturing of males and females.

The specificity of some types of population structures allows the following conclusions to be made:

- analysis of the properties of populations and various periods of their life shows that the composition of populations of various species in the same biocenosis, as well as the composition of the populations of one species in various biocenoses, are different;
- studying the composition of populations is important not only theoretically but also practically since variaged individuals in the populations differently react to the influence of various anthropogenous factors;
- the age structure of any population depends on the time of sexual maturity achievement, the total and average life expectancy of individuals, the reproduction period duration, the life expectancy of a generation, litter frequency, the character and dynamics of mortality in different sexual and age groups, and on fluctuations in the numbers of the population.

The causes to group populations within limited sites are various, but the main ones are in non-uniformity of the distribution of ecological conditions in geographical space and in similarity of the requirements of organisms of one species to these conditions. Functioning of a population in a biogeocenosis can be characterized by such more general parameters as the density of population, average biomass, and total abundance.

9.2. SPECIFICITY OF ANTHROPOGENIC INFLUENCE ON THE BIOSPHERE AND ITS STRUCTURAL ELEMENTS

The anthropogenous influence on the biosphere and its separate parts is characterized by the following regularities.

1. Every (without exception) ecosystem obeys the first law of thermodynamics. In essence, this law coincides with the basic law of nature: energy can neither arise nor disappear, but only pass from one form into another. This follows from the considered chain of energy transfer (transformation) through a trophic chain. According to this law, the total energy of the biosphere almost does not change.
2. The difference between the weight of substance entering the system and the weight of substance leaving it is equal to the weight of substance accumulated within the system. If no change (accumulation or decrease) occurs, then the system is in a steady state. In particular, for a chemical process: the weight of the initial reagents is equal to that of the final products. For biological systems such a state is referred to as *homeostatic*.
3. Every ecosystem obeys the second law of thermodynamics. Energy redistribution in any system (including ecological ones) is characterized by a quantity called *entropy*. At an equilibrium state (which any thermodynamic system aspires to) the entropy of the system takes on its maximum value. However, the process of photosynthesis is known to be accompanied by transformation of solar radiation energy into chemical one, which collects in substances and is equal to the total of all the chemical bond energies in molecules. Such a situation, on the first sight, contradicts the second law of thermodynamics since a higher-ordered form of matter organization appears and entropy should so decrease. This contradiction is due to that the second law of thermodynamics absolutely strictly applies to isolated, adiabatic or closed systems with no exchange processes with the environment. It is therefore possible to judge on the applicability of the second law of thermodynamics to organisms and, in general, to ecosystems, proceeding only from consideration of the total entropy change in the "ecosystem (organism)—environment" system: the sum of entropy differences at formation of organic substance ($\Delta S_1 < 0$) and in thermal processes proceeding in the environment ($\Delta S_2 > 0$). Then, $\Delta S = \Delta S_1 + \Delta S_2 > 0$, since $|\Delta S_2| >> |\Delta S_1|$, i.e. in an ecosystem participating in exchange processes with the environment (by substance, energy, and information) the entropy always increases, in full correspondence with the second law of thermodynamics.

4. All live substance in the biosphere has a common physicochemical basis, which also points to the unity of both inanimate and animate nature. The only distinction is only in the stability degree of natural objects (organism, population, and species) to external influences. Breaks in trophic chains result in the occurrence of poorly controllable sites, which reduces the stability of the ecosystem and biosphere as a whole. This causes the importance of biodiversity preservation and increase.
5. Substance, energy, information, and the physicochemical properties of ecosystems are interrelated so that any change of any of these parameters causes qualitative and quantitative changes in the structure, properties, and functions of the ecosystem. Therefore:
   - any change in the environment (substance, energy, the properties of a system) necessarily results in the occurrence and development of natural chain reactions to proceed towards either neutralization of this change or the formation of new natural systems;
   - the interaction of substance-energy components (gases, liquids, organisms) in the biosphere usually takes place under nonlinear laws; such processes are called synergetic, and the corresponding scientific lead is named synergetics; e.g., traces of sulfur or nitrogen oxides result in the formation of much sulfuric or nitric acids in the atmosphere, their precipitation onto the land or into reservoirs, and then to degradation of the soil cover, reservoir pollution, wood destruction, etc.;
   - all the abiotic and biotic components included in an ecosystem fit each other functionally; loss of one element of the system (e.g., disappearance of a species) necessarily leads to elimination of its other parts closely related to this element;
   - no ecosystem can steadily exist at artificial lack or surplus of one of the ecological elements; long changes in the quality and/or quantity of one of the components could result even in replacement of one system by another one.
6. Every ecosystem has specific properties and, consequently, the duration of passing of energy, substance and information fluxes for various systems is different. As an example it is possible to give such a parameter as the water exchange duration: few hours in an individual, ca. a week in the atmosphere, water renewal in rivers requires several weeks, in a lake—15–20 years, underground waters are renewed in ca. 1,500 years and the oceanic water—in 2,500 years.
7. For any ecosystems that direction of process development is always optimal which the minimum energy dissipation (or the minimum growth of entropy) corresponds to.
8. That system possesses the highest probability of self-preservation and self-regulation whose functions and properties correspond to effective input, processing and use of energy, substance, and information.
9. No part of an ecosystem can exist without other, functionally supplementing parts, and long functioning of organisms is only possible within the framework of ecosystems, which their components supplement each other in.
10. The efficiency of an ecosystem depends on the degree of interaction of its components and on the properties of the inhabiting organisms (their sizes, average life expectancy, reproduction rate, etc).
11. High-effective functioning of any system is only possible within certain spatial-temporary ranges. This means that the size of any system should correspond to its

functions, i.e. no system can decrease or extend unimpeded without damage to the efficiency of its functioning (e.g., any mammal cannot be smaller or bigger than those sizes which provide the best ability to reproduction).

12. The limiting factor of normal organism life is any factor whose value may be either minimal or maximal, and the range between them determines the tolerance (endurance) of a living organism to this factor.
13. At transition from one historical period to another one, obtaining a unit of useful production from natural systems requires more and more energy expenses. Simultaneously, the energy expenses per man grow. E.g., in the Stone Age the energy expense per man was ~4 thousand kcal per day, in the agrarian community — 12 thousand, in the industrial epoch — 70 thousand, now in the most advanced countries they are up to 250–300 thousand kcal/day. One of the consequences of this regularity is a so-called decreasing efficacy in agrosystems, i.e. an increase of the specific energy investment into an agrosystem corresponds to an increase of its productivity more and more poorly.
14. *Principle of emergence.* The sense of this principle is, first, in the presence of specific properties and attributes in the whole system which are absent in its parts, and, second, in the possible presence of a simple sum of elements not connected by special interrelations and interactions. Two illustrative examples are: i) any molecule has properties different in comparison with its separate constituent atoms, and simple mechanical summation of the atoms does not give the qualitative specificity of the molecule; ii) the presence of separated organs themselves does not allow getting a whole organism. The principle of emergence has great importance for the formation of correct systematic thinking and should be taken into account at implementation of ecological forecasting and ecological examination. In particular, at modeling of a biogeocenosis, a special quantity named the emergence factor is introduced to reflect the qualitative distinction between the model and original. The value of the emergence factor usually grows with increasing of the complexity of a natural system.
15. *Le Chatelier—Brown's principle.* At an external influence drawing a system from its state of steady equilibrium, the equilibrium is displaced towards that direction which the effect of the external influence weakens in. Practically, this means that the more deviation from ecological balance, the more energy expenses are required for attenuation of the influence of this external factor.
16. *Rule of moderate transformation of natural systems.* At exploitation of natural systems one should not exceed those limits which allow systems to keep their properties of adaptation, i.e. the ability to self-organization and self-regulation. This rule has several consequences.
    i) a unit of a renewable resource can be obtained only in an interval of time during which no limiting values of the used resources are exceeded;
    ii) nature protection measures (wood cultivation, reservoir purification, etc.) are possible to a certain (optimal) degree only;
    iii) artificial measures of influence (e.g., technical systems) are always less effective than natural systems or processes (even man-adjustable).
17. *Rule of "gentle" nature management.* Nature-conformable ("gentle") management of natural processes should be based on the use of objective laws of nature and

measures to promote favorable and long-term results, meet the present and future generations' interests.

18. *Rule of "severe" nature management.* Solely technical ("severe") management of natural processes inevitably leads to destructive social, ecological and economical consequences proceeding during a rather long period.
19. *One-percent rule.* A change in the energetics of a natural system beyond one percent of the total Earth intake of solar energy draws the system from its ecological balance. This is confirmed by experimental research of global photosynthesis, powerful cyclones, volcanic eruptions, and other similar processes characterized by their total energy not exceeding one percent of the solar radiation energy. Falling of the used energy amount of traditional resources (oil, gas, coal, nuclear fuel) outside a threshold of one percent frequently results in essential anomalies: sharp climatic changes, the hotbed effect amplification, occurrence of large wood and steppe fires.
20. *Ten-percents rule.* Transition from one link of a trophic chain to another one is accompanied by conservation of 10% of the energy, which corresponds to optimal functioning of the system.

9.3. Some Concepts of Geoecology

Vernadsky's works on geochemistry and the aggravation of the ecological crisis in the lithosphere which possesses a certain specifics are preconditions of the emergence of this scientific lead. First, the lithosphere is a substratum of development of the biosphere and human activity. Second, the lithosphere is the final sphere of burial of the activity products of live substance and the man. Third, processes in the lithosphere proceed rather slowly, which makes an erroneous impression of their insignificant ecological importance. Fourth, the consequences of any impact upon the lithosphere in many cases can have catastrophic character for the man (technogenic earthquakes, landslide activization, etc.)

Underground nuclear explosions, formation of solid waste dumps, pollution of the underground waters, impacts on the top horizons (thermal, electromagnetic, etc.), promptly increasing mining are considered as main types of the anthropogenous negative impact upon the lithosphere. The rate of mechanical substance extraction from the lithosphere is above 100 billion ton/year, which more than four times exceeds the denudation drift of all flowing waters on the terrestrial surface into reservoirs. All flowing waters move about 13 km^3/year of solid substance over the terrestrial surface. This is by 30 times less than the solid substance moved in the course of building works and mining. The production capacity nearly doubles every 15 years. Hence, the activization of the adverse (for the man) geological processes (landslides, mud flows, flooding and bogging, soil salinization, technogenic earthquakes, etc.) becomes clear. Taking into account agricultural objects, the man has changed the appearance of about half of the land area. The man's mechanical penetration deep into the lithosphere (underground tunnels, subways, pedestrian subways, storages, wells) promptly develops as well. The total length of the railways (and, therefore, the zones of changed soils, relief, geological processes) is about 1,400 thousand km, which by 3.5 times exceeds the Earth-to-Moon distance. The length of the main and ship canals is close to the Earth-to-Moon distance. As a result of the geological activity of mankind, so-called technogenic deposits develop.

Their geological characteristics are determined by the features of structure of the root rocks undergone transformation and the specifics of the corresponding anthropogenous process. Technogenic (anthropogenous) elements and relief forms are also classified as positive (embankments, waste heaps, dumps) and negative (excavations, pits, ditches).

In these conditions, the increasing importance is gained by new scientific leads which study versions of the ecological functions of the lithosphere (resource, geodynamic, geochemical, and geophysical ones). As a result, a new adjacent science, namely, *palaeoecology*, has appeared to study the biocenoses of the remote geological past.

The interrelation of historical geoecology and palaeoecology with other sciences causes inevitable analysis of the global ecological crises in the history of Earth, in particular, the causes of mass extinctions in the geological past (paleontology) in certain physic-geographical conditions (paleogeography) with subsequent comparison with the causes of extinction of the modern forms of organisms (biology) in certain physicogeographical conditions (geography) and/or under anthropogenous influence (ecology). Such crossing of the past and the present is necessary for more complete analysis of the essence of the phenomena occurred and occurring in the extinct, dying-out, and existing communities of organisms, and for forecasting mankind's actions reducing the consequences of crisis situations. In recent years, a number of hypotheses have been offered to try to schematically explain the mass extinction of organisms between the Mesozoic and Cenozoic Eras.

Hypothesis 1. Climate changes caused death of the plants eaten by herbivorous dinosaurs, and no new elements of flora to satisfy them as food appeared. With reduction of the numbers of the herbivorous forms, the quantity of predatory reptiles decreased.

Hypothesis 2. Orogeny changed the relief and led to reduction of the areas occupied by bogs, oases, and lagoons, i.e. the optimum places for reptile habitation. This could lead to reduction of the number of dinosaurs but not to their complete extinction.

Hypothesis 3. Diseases, degeneration, and parasites mortally tortured dinosaurs. Facts are known of finding dinosaur eggs with a very thick shell which the cub could not get out independently from.

Hypothesis 4. Falling of a giant meteorite, its explosion with the subsequent effect of "nuclear winter". Meteoric rains, active volcanic activity, changes in atmospheric pressure, passing of a huge comet near our planet, an increased intensity of solar radiation. However, it is not clear why in these catastrophic conditions only dinosaurs were lost but no other groups of animals and plants died out?

Hypothesis 5. Mammals ate all the eggs of dinosaurs, predatory dinosaurs ate all herbivorous ones, and then they ate each other.

Hypothesis 6. In the history of Earth, repeated inversions of its magnetic field with shifts of the geomagnetic poles took place. The animals regularly migrating from some places to others lost orientation and perished in whole flocks.

Hypothesis 7. Meteorites are known to drop onto Earth all the time, which leads to increase of the mass of our planet and, therefore, to increased gravity. Giant herbivorous forms lost their ability to movement since their skeleton did not sustain their weight (dinosaurs crushed themselves under the influence of the increased force of terrestrial gravitation).

Hypothesis 8. Some aliens having arrived to Earth exterminated all dinosaurs. In confirmation of this hypothesis it is possible to present pangolin skulls found in the USA with roundish holes between their eyes.

Thus, this short review of the hypotheses concludes that the existence of any species depends on many geoecological factors.

9.4. Concept of Sustainable Development

The beginning of the third millennium coincided with the need of mankind's transition to a new form of its development called as *sustainable development*. The inevitability of the transition of any country on our planet to sustainable development follows from the need of solving global problems common for the world community.

A new civilizational model has arisen as an attempt to find the most general concept of joint survival of the population on our planet, a way out from the global (and especially ecological) crisis state, prevention of the world accident—omnicide (death of all the alive). As the basic problems of ecological character of the 21st century are considered: climate change as a result of hotbed gas emission, deficit and pollution of fresh water, wood disappearance and desertification, biodiversity reduction, population growth (and migration), waste disposal, air pollution, degradation of soil and ecosystems, chemical pollution, exhaustion of the ozone layer, urbanization, the appreciable reduction of natural resources, breakages of biogeochemical cycles, spreading of diseases (including the emergence of new ones), etc. Mankind has faced contradictions between its growing needs and the inability of the biosphere to meet them without destroying the environment. The ideas of sustainable development meet the objective requirement of time and can affect, in a solving way, the future of each country of the world community, play an important role in the arrangements of state priorities, the social and economic development strategy of each country.

The notion of "sustainable development" was widely adopted after publication of the report prepared for the United Nations in 1987 by the World Commission on Environment and Development (WCED) headed by Gro Harlem Brundtland, the former Prime Minister of Norway.

The sense of this notion consists in that *sustainable development* is such a kind of development which meets the needs of the present without compromising the ability of future generations to meet their own needs.

Sustainable development can be recognized with the help of anthropocentric and biospherocentric indicators. The *anthropocentric* indicator in a broad sense is understood as survival of mankind (or a certain country) and the ability (possibility) of its further continuously long development. Our descendants should not have less possibility in comparison with the present generations to meet their needs. The *biospherocentric* (generally—ecological) indicator is related with the biosphere (the environment) preservation as the natural basis of the life on Earth (and in space) in order that further development of mankind would not proceed in an ecophobic form. Therefore, sustainable development is only possible with observance of the following basic principles:

- every person has the right to his/her healthy and fruitful life in harmony with nature, to life in the favorable environment;
- social and economic development should be directed to improvement of the life quality of people within acceptable limits of the economic capacity of ecosystems;
- development should go not harmfully to the environment but provide the possibility to meet the basic vital needs of both present and future people generations;
- preservation of the natural environment should be an integral part of the process of sustainable development; economic development, social justice, and ecological safety (which in total determine the main criteria of development) should be joined into a single whole;
- survival of mankind and stable social and economic development should be based on regularities of biotic regulation with biodiversity to be preserved;
- rational environmental management should be based on the optimum usage of both renewable and irrenewable resources, utilization and safe burial of wastes;
- ecologically-safe economy should be based on strengthening of the interrelation between economy and ecology, the formation of a uniform (conjugated) ecologized economic system of development;
- implementation of an appropriate demographic policy should be aimed at stabilization of the population and optimization of the human activity scales according to the fundamental laws of nature;
- the wide usage of the principle of anticipation is necessary, i.e. undertaking effective measures to prevent deterioration of the environment condition, prevention of ecological and technogenic catastrophes;
- prevention of big distinctions in the living standard of people is an important condition;
- usage of a variety of ownership forms and the mechanism of market relations should be focused on harmonization of the social relations when ensuring public safety;
- the importance of questions of rationalization of the scales and structure of personal consumption of the population in the long term should increase;
- preservation of the small people and ethnoses, their cultures, traditions, and habitats should be a priority of the state policy;
- development of the international cooperation and global partnership with a view of preservation, protection, and restoration of the integrity of the global ecosystem should be directly related with the acceptance of relevant international agreements and other legal acts by countries;
- free access to ecological information and the design of the corresponding database with the usage of global and national communications and other means of informatics in these purposes are necessary;
- in the course of legislative base development it is necessary to consider the ecological consequences of prospective actions, to proceed from the need of increased responsibility for ecological offenses, to provide compensations to those affected by environmental pollution;

- advance of the intellectual and cultural values onto a priority place relative to the material ones should promoted by ecologization of the human consciousness and outlook;
- the sovereign rights of every state to develop its own natural resources should be realized with no harm to the ecosystems outside of its frontiers; recognition of the principle of differentiated responsibility of countries for disturbance of global ecosystems is important in international law.

9.5. Ecologization of Science and Engineering: Interdisciplinary Links and Interactions

At various historical stages of the development of natural sciences, these or those tendencies acted were dominant. In particular, the development of physics, biology or space exploration had an essential impact upon the orientation and character of natural and technical sciences and the humanities. In the conditions of "the environmental twenty years" (1972–1992) the determinant character of the man—nature, society—biosphere relationship problem in the dynamics of world-wide and historical development of the civilization caused the formation of a new tendency—science ecologization as penetration of ecological factors into the developed system of natural and technical sciences and the humanities. Three levels of this ecologization are counted:

1) *intradisciplinary ecologization*: revealing biospheric knowledge in the system of a specific scientific discipline;
2) *interdisciplinary ecologization*: the formation of new ecologically-orientated disciplines (in natural sciences: eco-mathematics, eco-physics, eco-chemistry, etc.; in technical sciences: eco-engineering, eco-cybernetics, eco-systematics, etc.; in the humanities: eco-economics, eco-law, eco-psychology, etc.) within the traditional system of sciences;
3) *problematic ecologization*: the application of various fields of natural-scientific knowledge to solve specific ecological tasks.

Features of Eco-Physics

In the first half of the 20th century the physical picture of the world dominated in the course of interpretation of objective reality. The physical and mathematical methods of research actively extended to the fields of other sciences. In the second half of the same century, at formation of an objective picture of the world, absolutization of physical and mathematical representations lost its traditional heuristic appeal to a certain degree.

The internal logic of the development of modern physics allows drawing a conclusion that the expected qualitative leap in physical reality cognition can give a new impulse to theoretical views and conceptual constructions. E.g., in the mid-nineties the physics of high energies went on the level of new concepts (anti-atom, anti-proton, etc.), which considerably supplemented the traditional physical picture of the world. Besides, in physics, in particular, there arise situations whose overcoming requires new approaches, e.g., more fundamental usage of biological and ecological representations at the formation of new physical views. Under the direct influence of biological ideas in analysis of processes in the microworld the characteristics corresponding to spontaneously developing systems are applied. This means the need of some modification of physics when extending its concepts to animate natural processes. The ecologization of modern physical knowledge is most affected by some external factors being function of specific conditions (social, economic, culturological ones, etc.). On the one hand, it is expressed in the requirement of "practization" of the corresponding physical research and development. Their development (and respectively—financing) is increasingly associated with solution of mankind's vital problems. On the other hand, the genesis of negative (for the man and his environment) social and ecological consequences is associated with the development of physical knowledge. World experience gives evidence of that more and more material, financial and intellectual resources are invested into orientation of physical research and development towards the directions to provide a high level of safety of nuclear power stations and other objects of the technosphere, effective utilization of radioactive wastes, etc.

Ecological ideas, notions, and concepts thus get into the system of physical and mathematical knowledge to some extent, especially into its applied spheres, both as proceeding from an "external" need caused by social requirements and according to the internal logic of the development of natural sciences. Just in this context, the ideas of "ecological physics" make sense, in the course of whose formation, conceptual representations of physical knowledge extend to analysis of the biospheric phenomena and processes, i.e. the regularities of relationship among living organisms (and their communities) with the environment are studied by means of physical and mathematical methods.

From Chemistry to Eco-Chemistry

Chemical processes constitute one of the bases of the practical attitude to natural substance. Scientific chemicalization is understood as the active usage of the laws of chemistry to intensify substance exchange in production dynamics. At the same time, large-scale chemicalization leads to that the increasing quantity of substances and materials inacceptable for the biota (xenobiotics) gets into historically developed ecosystems. The degree and scales of chemical impact on natural processes and relations approach such a level that the traditional integrity and stability of the biosphere appear under threat. The chemical knowledge and realization of its practical results increasingly manifest themselves, first, in strengthening of the degradation changes of the native human environment under the influence of emissions of technical and anthropogenous character, and, second, in the development of means, methods, and specifications providing the development of production activity according to modern social and ecological representations.

The development of ecological chemistry (or chemical ecology) is associated with the development of both general and specific directions of solution of the specified biospheric contradictions. E.g., inclusion of polymeric materials into the native environment has promoted intense development in the field of synthesis of "disappearing plastics" as substances decomposing under the influence of external factors (water, light, etc.) into rather safe components. Pesticide evolution is also indicative. If their first generations are connected with intense accumulation of negative factors for the man and his environment, the subsequent modifications, possessing an increased selectivity and various duration of influence, more organically blend with biospheric processes. Similar modification is also characteristic for chlorofluorocarbons (Freon™) participating in destruction of the atmospheric ozone layer of Earth. In the conditions of strengthening of degradation changes of the native environment mainly caused considerably by the increase of emissions of substances of technogenic origin into the biosphere, and also as a result of aggravation of the problem of deficiency of natural resources, the waste regeneration need has amplified. Orientation to the design of rather closed technologies (dominating use of recirculation principles) has certain advantages. Much bigger prospects are opened before biotechnologies on the basis of which the interrelation of chemical and biological (microbiological) processes is carried out and which, more or less, correspond to the biospheric links and regularities.

Priority of Biology

In the mid-20th century, important discoveries occurred connected with radical transformation of the ideas of specifics of biological objects, formation of the base of modern molecular and genetic biology.

The progress of biological sciences promoted revealing of the development regularities of animate nature, i.e. more fundamental inclusion of biological knowledge into the generalized scientific picture of the world. The modern biological research is not limited to an analytical approach but is guided by the representations of integral character corresponding to the specifics of biological objects. In the context of the social and ecological situation, biological research gains integrated character and biospheric scale, especially considering their special role in solution of most actual problems of the present (food, demographic, ecological ones). Owing to the methods of genetic engineering it becomes real to purposefully change the historically developed organic forms, which is important for increasing the productivity of traditional agroecosystems, enhancing the food resources of mankind, etc. A strong base is laid under modern medicine, genetic and cellular engineering, immunology, microbiological synthesis, underlying medico-biological research and providing effective treatment of genetic diseases, malignant tumors, etc. Thereby, preconditions for preservation and increase of biological diversity are created.

Analysis of "external" and "internal" factors of the development of the biological knowledge system allows stating the fact of its known ontological (practical) advantage over physics, having, besides other, a social and ecological context. First, just traditional biology includes general ecology in its structure.

Biology is one of few disciplines of modern scientific knowledge which possesses historical experience of analysis of phenomena and processes in the system of relationships of a living organism and its environment. Second, the biological sciences, assimilating and developing the research principles of physical and mathematical areas of knowledge, have revealed both their universality and a certain limitation in the interpretation of some phenomena and processes of biogenic character.

Inclusion of ecological representations raises the status and importance of the integrative approach in natural sciences, allowing determining the possibilities and restrictions of the reduction style of thinking more accurately. Third, the biological sciences show a high practicality in relation to the corresponding results of basic physical research and development. For public consciousness the development of physics is associated with the real tragic consequences of Chernobyl, and the achievement of biology is with actual positive or negative potential facts. Fourth, the biological sciences have traditional humanistic orientation, which considerably raises their status in the conditions of strengthening of "dehumanization" of physics and the natural sciences in general.

At the same time, all this does not mean the existence of opposition of the system of physical and biological sciences at all. Moreover, only with their effective interaction and interrelation it is possible to rely on further revealing of fundamental natural regularities concerning the "man—biosphere" system. Therefore, in conditions when the natural sciences are guided by solution of environmental problems, the physical and biological knowledge is drawn nearer further.

When the contradictions in the "man—society—biosphere" system are eliminated, on the one hand, the orientation of physics towards solution of global problems of the present, including the ecological one, becomes stronger. On the other hand, the level of "theorization" of biology providing a further increment of knowledge in the field of living matter rises.

Role of the Earth Sciences

Among the natural-scientific disciplines which have had special stimulus of development under the influence of the modern ecological situation, it is necessary to consider the Earth sciences. Unfortunately, in the hierarchy of modern scientific knowledge the disciplines relating to the Earth sciences are traditionally considered as outsiders. No ideas of basic research and development capable (as is accepted in relation to physics, chemistry or biology) of rendering radical changes in the existing picture of the world are usually associated with them.

Geography and geology were mainly guided by describing and classifying spatial natural systems. Their temporary evolution was not considered in a due measure. This circumstance probably explains that historic fact that during the era of great geographical discoveries of the 15—17th centuries (when there appeared a problem of spatial studying of natural systems) just geography as the leader of the Earth sciences represented itself as one of the leading disciplines of the natural-scientific cycle.

By the beginning of the 20th century, the sciences of the leading direction (say, physics), expanding their theoretical bases, fixed the scientific status in the formation of an objective picture of the world. The disciplines similar to the Earth sciences quite long time remained mainly at a descriptive level.

At the same time, assimilation of theoretical methods and representations developed within the general natural sciences had fundamental value for the development of the Earth sciences, promoting the formation of the corresponding perspective leads (geophysics, geoecology, mathematical geography, geochemistry, etc.).

Now it is even accepted to consider that the Earth sciences have appeared better prepared for comprehension of contradictions in the "biosphere—man" system. The opportunities provided by the modern Earth sciences for successful solution of many social environmental problems push them into the group of leading disciplines of modern scientific knowledge.

Natural Sciences and Engineering Interrelation

The ecological principles, rules and laws, actively penetrating into the structure of natural-scientific knowledge, have an essential impact on the nature of development of technical sciences. The interrelation of the components of the “natural sciences—technical sciences” system is caused by that the results of both fundamental and applied research provide more and more perfect methods of materialization of the revealed natural regularities to engineering. For the said system characteristic are such two interconnected processes as “scientification” of technical knowledge (i.e. engineering development on the basis of introduction of the results of basic natural-scientific research) and “technologization” of natural-scientific knowledge (i.e. creation of technico-technologically focused theories). Just on the joint between the natural and technical disciplines there is progress in the scientific and technical sphere (microelectronics, biotechnologies, etc.). The development of technical and ecological knowledge and the phenomenon of its materialization have the increasing impact on both immanent processes and socio-natural systems. Such an influence historically has both negative and positive orientation, which manifests itself as the ratio of different concepts in scientific and technical development, namely, technological optimism (technophilia) and technological pessimism (technophobia). *Technological optimism* is an outlook within which the positive prospects of the development of engineering and technologies are absolutized. *Technological pessimism*, on the contrary, focuses attention on the negative moments of technical and technological development of the civilization. Depending on various objective and subjective factors in public consciousness, one or the other outlook dominates. In the second half of the 20th century and at the beginning of the 21st century the concept according to which just the technology is considered as a “key” to positive resolution of conflicts in the “man—biosphere” system still prevails.

9.6. Concepts of the Chemical Physics of Polymers in Technologies and Environment Protection

Chemical physics is a most important fundamental science in the modern natural sciences. The concepts of chemical physics qualitatively and quantitatively explain the mechanisms of various processes, such as the reactions of oxidation, burning and explosion, obtaining food products and drugs, oil hydrocarbon cracking, making polymeric materials, biochemical reactions underlying metabolism and genetic information transfer, etc. Chemical physics comprises notions of the structure, properties, and reactivity of various substances, free radicals as active centers of chain processes proceeding in both mineral and organic nature, the scientific foundations of low-waste and resource-saving technologies. The design and wide usage of synthetic polymeric materials is a lead in chemistry. This results in the emergence of new environmental problems due to pollution of the environment with these materials and the wastes of their production and monomer synthesis. The presented below results of our studies of the kinetics and mechanism of polymerization of vinyl monomers which can proceed in quite various conditions point to their usability in the scientific justification of optimization of technological modes of monomer and polymer synthesis, and at solving both local and global environmental problems.

To obtain reliable experimental data and to correctly interpret them, we used such physicochemical and analytical techniques as dilatometry, viscometry, UV and IR spectroscopy, electronic paramagnetic resonance, light scattering spectroscopy, electron microscopy, and gas-liquid chromatography. To analyze the properties of polymeric dispersions, the turbidity spectrum method was used, and the efficiency of flocculants was estimated gravimetrically and by the sedimentation speed of special suspended imitators (e.g. copper oxide).

Effect of salts of metals of varying valency. Additives of the stearates of iron (IS), copper (CpS), cobalt (CbS), zinc (ZS), and lead (LS) within a certain concentration range were found to increase the polymerization rate of styrene and methylmethacrylate (MMA) in comparison with their thermopolymerization. By initiating activity, they can be arranged as LS < CbS < ZS < IS < CpS. The decrease in the values of the effective activation energy, the activation energy of reaction initiation, and the kinetic reaction order by monomer point to the active participation of the monomer in chain initiation. IR spectroscopy shows that an intermediate monomer–stearate complex is formed which then decomposes into active radicals to initiate polymerization. The benzoyl peroxide (BP)–IS (or CpS) systems can be used for effective polymerization initiation. A concentration inversion of the catalytic properties of stearates has been found, which depends on salt concentration and conversion degree. The efficiency of the accelerating influence decreases with increasing temperature, the BP–IS system possesses the highest initiating activity. The initiating mechanism for these systems is principally different that the redox one. It follows from experimental data (color changes before and in the course of polymerization, the absorption and IR spectra of reactive mixtures, electron microscopy observations, etc.) that initiation occurs due to stearate radicals formed at decomposition of the complex consisting of one BP molecule and two stearate ones.

The phenomenon of concentration and temperature inversion of the catalytic properties of gold, platinum, osmium and palladium chlorides at thermal and initiated polymerization of styrene and MMA has been discovered. The mechanism of ambiguous action of noble metal salts is caused by the competition of the initiating influence of monomer complexes with colloidal metal particles and the inhibition reaction proceeding by ligand transfer.

Effect of some organic solvents. The influence of acetonitrile (ACN) and dimethyl formamide (DMFA) on the radical polymerization of styrene and (met)acrylic esters initiated with azo isobutyronitrile (AIBN) or BP was studied. Basic characteristics of the gross kinetics of polymerization in solution have been found. The reaction order by initiator concentrations is always 0.5, which points to bimolecular chain termination. The reaction order by monomer concentration varies within rather wide limits (above or below unity) depending on the chemical nature of the solvent, monomer, and initiator. The lowest (0.83) and highest (1.6) value were found for BP-initiated polymerization of MMA in DMFA solution and for AIBN-initiated polymerization of styrene in CAN solution, respectively. Such a high value of the order by monomer is caused by the abnormally low rate of AIBN decomposition in ACN, which, in turn, is explained by the donor-acceptor interaction of the alkyl and nitrile group of the initiator with the nitrile and alkyl group of the same solvent molecule. The influence of DMFA and CAN on all the elementary stages of polymerization has also been ascertained, which manifests itself in the dependences of the rate constants of the reactions of initiation, propagation, and termination on the monomer concentration. This is caused by such factors as changes in the initiator decomposition rate, macroradical solvation with the molecules of an electron-donor solvent, diffusional-controlled chain termination, and conformational changes of macromolecules in solution.

Homopolymerization and copolymerization of acrylonitrile in an aqueous solution of sodium sulfocyanide. When acrylonitrile (AN)-based fiber-forming polymers are obtained, a spinning solution ready for fiber formation appears as a result of polymerization in some solvent. Organic solvents or solutions of inorganic salts are used as solvents in these cases. First, a comparative study was made of the kinetics and mechanism of polymerization of AN in DMFA and in an aqueous solution of sodium sulfocyanide (ASSSC) initiated with AIBN and some newly-synthesized azo nitriles (azobiscyanopentanol, azobiscyanovalerial acid, azobisdimethylethylamidoxime). The polymerization rate in ASSSC turns out to be significantly higher in comparison with the reaction in DMFA, in spite of the initiation rate in the presence of the said azo nitriles being more than 1.5 times lower than that in DMFA. The lower initiation rate in ASSSC is associated with stronger manifestation of the "cell effect" due to the higher viscosity of the water–salt solvent and the ability of water to form H-bonds (which hinders initiation). The ratio of the rate constants of chain propagation and termination ($K_p/K_t^{0.5}$) was found to be ca. tenfold higher in ASSSC than in DMFA. The molecular mass of the polymer formed is correspondingly higher. These differences are caused by the influence of medium viscosity on K_t and formation of H-bonds with the nitrile groups of the end chain of a macroradical and the added monomer. But sodium sulfocyanide (which forms charge-transfer complexes with a molecule of AN or its radical to activate them) mainly contributes into increasing K_p.

The kinetics of AN copolymerization with methylacrylate (MA) or vinylacetate (VA) in ASSSC is qualitatively analogous to AN homopolymerization in identical conditions. At the same time, the initial reaction rate, copolymer molecular mass, effective activation energy (E_{ef}), orders by initiator and total monomer concentration differ. E.g., the decrease in E_{ef} is due to the presence of a more reactive monomer (MA), and in the AN–VA system the non-end monomer chains in macroradicals influence the rate constant of cross chain termination. For these binary systems, the copolymerization constants were estimated, whose values point to a certain mechanism of chain growth, which leads to the MA concentration in the copolymer being significantly higher than in the source mixture. The same is observed in the case of AN with VA polymerization (naturally, the absolute amount of AN in the final product is much higher than MA or VA due to its higher initial concentration in the mixture).

Obtaining synthetic PAN fiber of a nitrone type is preceded by the formation of a spinning solution by means of copolymerization of AN with MA (or VA) and itaconic acid (IA) or acrylic acid (AA) or methacrylic acid (MAA) or methallyl sulfonate (MAS). Usually, the mixtures contain 15% of AN, 5–6% of the second monomer, 1–2% of the third one, and ca. 80 wt. % of 51.5% ASSSC. When a third monomer is introduced into the reaction mixture, the rate of the process and the molecular mass of the copolymer decrease, which makes these comonomers be peculiar low-effective inhibitors. In such a case, it becomes possible to estimate the inhibition (retardation) constant which is, in essence, the rate constant of one of the reactions of chain propagation. At copolymerization of ternary monomeric systems based on AN in ASSSC, the chain origination rate increases with the total monomer concentration. The initiation reaction order relative to the total monomer concentration varies from 0.5 to 0.8 depending on the degree of the retarding effect of the third monomer. Besides, AN in the three-component system is shown to participate less actively in the chain origination reaction than at its homopolymerization and copolymerization with MA or VA.

Thus, the obtained results enable regulating the copolymerization kinetics and the structure of the copolymer formed, which, finally, is a way of chemical modification of synthetic fibers.

Stable radicals in the polymerization kinetics of vinyl monomers. Free radicals are neutral or charged particles with one or more uncoupled electrons. Unlike usual (short-living) radicals, stable ones (long-living) are characteristic of paramagnetic substances whose chemical particles possess strong delocalized uncoupled electrons and sterically screened reactivity centers. This is the very cause of the high stability of many classes of nitroxyl radicals of aromatic, fatty-aromatic and heterocyclic series, and ion radicals and their complexes.

Peculiarities of thermal and initiated polymerization of vinyl monomers in the presence of anion radicals of tetracyanoquinodimethane (TCQM) were investigated. TCQM anion radicals are shown to effectively inhibit both thermal and AIBN-initiated polymerization of styrene, MMA, and methylacrylate in acetonitrile and dimethylformamide solutions. Inhibition is accomplished by the recombination mechanism and by electron transfer to the primary (relative to the initiator) or polymeric radical. The electron-transfer reaction leads to the appearance of a neutral TCQM which regenerates the inhibitor in the medium of electron-donor solvents. Our calculation of the corresponding radical-chain scheme has allowed us to derive an equation to describe the dependence of the induction period duration on the initiator and inhibitor concentrations, and how the polymerization rate changes with time. The mechanism of the initiating effect of the peroxide—TCQM system has been ascertained, according to which a single-electron transfer reaction proceeds between an anion radical and a BP molecule, with subsequent reactions between the formed neutral TCQM and benzoate anion, and a benzoate radical and one more anion radical TCQM. Free radicals initiating polymerization are formed at the redox interaction between the products of the said processes and peroxide molecules. The TCQM anion radical interacts with peroxide only at a rather high affinity of this peroxide to electron (BP, lauryl peroxide); in the presence of cumyl peroxide, the anion radical inhibits polymerization only.

Iminoxyl radicals which are stable in air and are easily synthesized in chemically pure state (mainly, crystalline brightly-colored substances) present a principally new type of nitroxyl paramagnets. Organic paramagnets are used to intensify chemical processes, to increase the selectivity of catalytic systems, to improve the quality of production (anaerobic hermetics, epoxy resins, and polyolefins). They have found application in biophysical and molecular-biological studies as spin labels and probes, in forensic medical diagnostics, analytical chemistry, to improve the adhesion of polymeric coatings, at making cinema and photo materials, in device building, in oil-extracting geophysics and defectoscopy of solids, as effective inhibitors of polymerization, thermal and light oxidation of various materials, including polymers.

In this connection, systematic studies were made of the inhibiting effect of many stable mono- and polyradicals on the kinetics and mechanism of vinyl monomer polymerization. The efficiency of nitroxyls as free radical acceptors has promoted their usage to explore the mechanism of polymerization by inhibition. Usually, nitroxyls have time to react only with a part of the radicals formed at azo nitrile decomposition, and they do not react at all with primary radicals at peroxide initiation. Iminoxyls have been found to terminate chains by both recombination and disproportionation in the presence of azo nitriles. Inhibitor regeneration proceeds as a result of detachment of a hydrogen atom from an iminoxyl by an active radical to form the corresponding nitroso compound. The mechanism of inhibition by a nitroso compound is addition of a growing chain to a -N=O fragment to form a stable radical again. The interaction of iminoxyls with peroxides depends on the solvent type. E.g., in vinyl monomers, induced decomposition of benzoyl peroxide occurs to form a heterocyclic oxide (nitrone) and benzoic acid. In contrast to iminoxyls, aromatic nitroxyls in a monomeric medium interact with peroxides to form non-radical products. Imidazoline-based nitroxyl radicals possess advantages over common azotoxides which are their stability in acidic media (owing to the presence of an imin or nitrone functional groups) and the possibility of complex formation and cyclometalling with no radical center involved.

Monomer stabilization. The practical importance of inhibitors is often associated with their usage for monomer stabilization and preventing various spontaneous and undesirable polymerization processes. In industrial conditions, polymerization may proceed in the presence of air oxygen and, hence, peroxide radicals MOO$^{\cdot}$ serve active centers of this chain reaction. In such cases, compounds with mobile hydrogen atoms, e.g. phenols and aromatic amines, are used for monomer stabilization. They inhibit polymerization in the presence of oxygen only, i.e. they are antioxidants. As inhibitors of polymerization of (met)acrylates proceeding in the atmosphere of air, some aromatic amines known as polymer stabilizers were studied, namely, dimethyldi-(*n*-phenyl-aminophenoxy)silane, dimethyldi-(*n*-β-naphthylaminophenoxy)silane, 2-oxy-1,3-di-(*n*-phenylaminophenoxy)propane, 2-oxy-1,3-di-(*n*-β-naphthylaminophenoxy)propane. These compounds have proven to be much more effective stabilizers in comparison with the widely used hydroquinone (HQ), which is evidenced by high values of the stoichiometric inhibition coefficients (by 3–5 times higher than that of HQ). It has been found that inhibition of thermal polymerization of the esters of acrylic and methacrylic acids at relatively high temperatures (100°C and higher) is characterized by a sharp increase in the induction periods when some critical concentration of the inhibitor $[X]_{cr}$ is exceeded. This is caused by that at polymerization in air, the formation of polymeric peroxides as a result of copolymerization of the monomers with oxygen should be taken into account. Decomposition of polyperoxides occurs during the induction period as well and can be regarded as degenerated branching. The presence of critical phenomena is characteristic of chain branched reactions. However, in early works describing inhibition of thermooxidative polymerization, no degenerated chain branching on polymeric peroxides was taken into account. It follows from the results obtained that the value of critical inhibitor concentration $[X]_{cr}$ can be one of the basic characteristics of its efficacy.

Inhibition of spontaneous polymerization of (meth)acrylates is necessary not only at their storage but also in the conditions of their synthesis which proceeds in the presence of sulfuric acid. In this case, monomer stabilization is more urgent, since sulfuric acid not only deactivates many inhibitors but is also capable of intensifying the process of polymer formation. The concentration dependence of induction periods in these conditions has a brightly expressed nonlinear character. Unlike polymerization in bulk, in the presence of sulfuric acid, decomposition of polymeric peroxides is observed at relatively low temperatures, and the values $[X]_{cr}$ for the amines studied are by ca. 10 times lower than $[HQ]_{cr}$.

Synthesis of MMA from acetonecyanhydrine is a widely spread technique of its industrial synthesis. The process proceeds in the presence of sulfuric acid in several stages, when various monomers are formed and interconverted. Separate stages of this synthesis were modeled with reaction systems containing, along with MMA, methacrylamide and methacrylic acid, and water and sulfuric acid in various ratios. As heterogeneous and homogeneous systems appeared at this, inhibition was studied in both static and dynamic conditions. The aforesaid aromatic amines appear to effectively suppress polymerization at different stages of the synthesis and purification of MMA. Their advantages over hydroquinone are strongly exhibited in the presence of sulfuric acid in homogeneous conditions, or under stirring in biphasic reaction systems. Besides, application of polymerization inhibitors is highly needed in dynamic conditions at the stage of esterification.

The usage of monomer stabilizers to prevent various spontaneous polymerization processes implies further release of the monomer from the inhibitor prior to its processing into a polymer. It is usually achieved by monomer rectification, often with preliminary extraction or chemical deactivation of the inhibitor, which requires high energy expenditures and entails large monomer losses and extra pollution of the environment. It would be optimal to develop such a way of stabilization where the inhibitor would effectively suppress polymerization at monomer storage but would almost not affect it at polymer synthesis. The usage of inhibitors low-soluble in the monomer is one of possible variants. When the monomer is stored and the rate of polymerization initiation is low, the quantity of the inhibitor dissolved could be enough for stabilization. Besides, as the inhibitor is spent, its permanent replenishment is possible due to additional dissolution of the earlier unsolved substance. The ammonium salt of N-nitroso-N-phenylhydroxylamine (cupferon), and some cupferonates were studied as such low-soluble inhibitors. The solubility of these compounds in acrylates, its dependence on the monomer moisture degree, the influence of the quantity of the inhibitor and the duration of its dissolution on subsequent polymerization were studied. Differences in the action of cupferonates are due to their solubility in monomers, their various stability in solution and the ability of deactivation; all this results in poorer influence of the inhibitor on monomer polymerization at producing polymer.

Some peculiarities of the kinetics and mechanism of emulsion polymerization. Emulsion polymerization, being one of the methods of polymer synthesis, enables the process to proceed with a high rate to form a polymer with a high molecular mass, high-concentrated latexes with a relatively low viscosity to be obtained, polymeric dispersions to be used at their processing without separation of the polymer from the reaction mixture, and the fire-resistance of the product to be significantly raised. At the same time, the kinetics and mechanism of polymerization in emulsion feature ambiguity, which is caused by such specific factors as the multiphasity of the reaction system and the variety of kinetic parameters whose values depend not so much on the reagent reactivity as on the character of their distribution over phases, reaction topochemistry, the way and mechanism of nucleation and stabilization of particles. The obtained results pointing to the discrepancy with classical concepts, can be characterized by the following effects: 1) recombination of radicals in an aqueous phase leading to a reduction of the number of particles and to the formation of surfactant oligomers capable of acting as emulsifiers; 2) the presence of several growing radicals in polymer-monomer particles, which causes the appearance of gel effect and the increase in the polymerization rate at high conversion degrees; 3) a decrease in the number of latex particles with the growth of conversion degree, which is associated with their flocculation at various polymerization stages; 4) an increase in the number of particles in the course of reaction when using monomer-soluble emulsifiers, and also due to the formation of an "own" emulsifier (oligomers).

Surfactants (emulsifiers of various chemical nature) are usually applied as stabilizers of disperse systems, they are rather stable, poorly destructed under the influence of natural factors, and contaminate the environment. The principal possibility to synthesize emulsifier-free latexes was shown. In the absence of emulsifier (but in emulsion polymerization conditions) with the usage of persulfate-type initiators (e.g., ammonium persulfate), the particles of acrylate latexes can be stabilized with ionized endgroups of macromolecules. The $^{\cdot}M_nSO_4^-$ ion radicals appearing in the aqueous phase of the reaction medium, having reached a critical chain length, precipitate to form primary particles, which flocculate up to the formation of aggregates with a charge density providing their stability. Besides, due to recombination of radicals, oligomeric molecules are formed in the aqueous phase, which possess properties of surfactants and are able to form micelle-like structures. Then, the monomer and oligomeric radicals are absorbed by these "micelles", where chains grow. In the absence of a specially introduced emulsifier, all basic kinetic regularities of emulsion polymerization are observed, and differences are concerned only with the stage of particle generation and the mechanism of their stabilization, which can be strengthened at copolymerization of hydrophobic monomers with highly hydrophilic comonomers. Increasing temperature results in the growth of the polymerization rate and the number of latex particles in the dispersion formed, decreasing their sizes and the quantity of the formed coagulum, and in improved stability of the dispersion. At emulsifier-free polymerization of alkyl acrylates, the stability of emulsions and obtained dispersions rises in the monomer row: methylacrylate < ethylacrylate < butylacrylate, i.e. the stability growth at lowering the polarity of the main monomer.

Our account of the aforesaid factors influencing the kinetics and mechanism of emulsion polymerization (in both presence and absence of an emulsifier) has enabled the influence of comonomers on the processes of formation of polymeric dispersions based on (meth)acrylates to be explained. Changes of some conditions of reaction have turned out to affect the character of influence of other ones. E.g., increasing the concentration of methacrylic acid (MAA) at its copolymerization with methylacrylate (MA) at a relatively low initiation rate leads to a decrease in the rate and particle number and to an increase in the coagulum amount. But at high initiation rates, the number of particles in the dispersion in the presence of MAA rises and their stability improves. The same effects were revealed for emulsifier-free polymerization of butylacrylate as well, when at high temperatures its partial replacement by MAA results in better stabilization of the dispersion, an increase in the reaction rate and the number of particles (whereas their decrease was observed in the presence of an emulsifier). Similar effects were found for AN as well, which worsens the stability of dispersion at relatively low temperatures but improves it at high ones. Increasing in the AN concentration in the ternary monomeric system with a high MAA content leads to a higher number of particles and better stability of dispersion at relatively low temperatures as well.

With the aim to explore the possibility to synthesize dispersions whose particles would contain reactive polymeric molecules with free multiple C=C bonds, emulsion copolymerization of acrylic monomers and unconjugated dienes was studied. The usage of such latexes to finish fabrics and some other materials promotes getting strong indelible coatings. The kinetics and mechanism of emulsion copolymerization of ethylacrylate (EA) and butylacrylate (BA) with allylacrylate (AlA) (with ammonium persulfate (APS) or the APS—sodium thiosulfate system as initiators) were studied. The found constants of copolymerization of AlA with EA ($r_{AlA} = 1.05$, $r_{EA} = 0.8$) and AlA with BA ($r_{AlA} = 1.1$, $r_{BA} =$

0.4) point to different degrees of the copolymer's unsaturation with AlA units. Emulsion copolymerization of multicomponent monomeric BA-based systems (with AN, MAA, and the unconjugated diene acryloxyethyl maleate (AOEM) as comonomers) was also studied. The influence of AN and MAA is described above. Copolymerization with AOEM depends on reaction conditions. E.g., in MAA-free systems, AOEM reduces the polymerization rate. In the presence of MAA, the rate of the process at high conversion degrees increases with the AOEM concentration, this monomer promoting the gel effect due to partial chain linking in polymer–monomer particles by the side groups with C=C bonds. The degree of unsaturation of diene units in the copolymer is subject to the composition of monomers, AOEM concentration, and temperature. In the BA—AOEM system, an increase of the diene concentration results in an increase in the unsaturation degree. This means that the diene radical is added to "its own" monomer with a higher rate than to BA ($r_{AOEM} = 7.7$). The higher unsaturation of diene units in MAA-containing systems points to that the AOEM˙ radical interacts with MAA with a higher rate than with BA, and the probability of cyclization reduces, the degree of polymer unsaturation increases, and the diene's retarding action upon polymerization weakens.

Compositions for production of rigid foamed polyurethane. In the field of polymer physicochemistry, studies were made according to the requirements of the Montreal Protocol (1987), which demands drastically reduce the production and consumption of chlorofluorocarbons (CFC, Freon™) and even replace oxone-dangerous substances by ozone-safe ones. Our investigations dealt with the replacement of trichlorofluoromethane (Freon-11), which had been used as a foam-maker in the synthesis of rigid foamed polyurethane (FPU) over a long period of time, which was thermal insulator in freezing chambers and building constructions. On the basis of our experimental dependences of the kinetic parameters of the foaming process (the instant of start, the time of structurization, and the instant of foam ending rising), the values of density and heat conductivity of pilot foamed plastics samples on the concentration of the reactive mixture components and physic-chemical conditions of FPU synthesis, optimal compositions (recipes) of mixtures with Freon-11 replaced by a ozone-safe (with an ozone-destruction potentials by an order of magnitude lower in comparison with Freon-11) azeotropic mixture of dichlorotrifluoroethane and dichlorofluoroethane were found. Their practical implementation requires no principal changes of known technological procedures and usage of new chemical reagents, which is an important merit of our developments.

Scientific basics of the synthesis of a high-molecular-weight flocculant. There exists a problem of purification of natural water and industrial sewage from various pollutants, including suspended and colloid-disperse particles, associated with the growing consumption of water and its deteriorating quality (owing to anthropogenic influence). It is known that flocculants can be used for these purposes, which are high-molecular-weight compounds capable of adsorption on disperse particles to form quickly sedimenting aggregates. Polyacrylamide (PAA) is most active. In connection with that many countries (including Russian Federation) suffer from acrylamide (AA) deficit, we have developed modifications of the synthesis of PAA flocculant by means of the usage of AN and sulfuric acid to implement the reactions of hydrolysis and polymerization. It has turned out that in the presence of a radical initiator of polymerization and sulfuric acid, AN participates in both processes simultaneously, and, as AA is formed from AN, their joint polymerization begins. Desired polymer properties were achieved at AN polymerization in an aqueous solution of sulfuric acid up to a certain conversion degree with subsequent hydrolysis of the polymerizate (a two-stage synthetic scheme) or at achieving on optimal ratio of the rates of these reactions proceeding in one stage. The influence of the nature and concentration of initiator, the content of sulfuric acid, temperature and reaction duration on the quantity and molecular mass of the polymer contained in the final product, its solubility in water and flocculating properties were studied. The required conversion degree at the first stage of synthesis by the two-stage scheme is determined by the concentrations of AN and the aqueous solution of sulfuric acid. At one-stage synthesis, changes in temperature and monomer amount almost equally affect the reaction rates of hydrolysis and polymerization and do not strongly affect the copolymer composition. But changes in the concentration of either acid or initiator rather strongly influence the molecular mass and composition of macromolecules, which causes extremal dependences of the flocculating activity on these factors.

Thus, on the basis of our above results, the following conclusions can be made.

1. New kinetic regularities have been revealed at polymerization of vinyl monomers in homophasic and heterophasic conditions in the presence of additives of transition metal salts, azo nitriles, peroxides, stable nitroxyl radicals and anion radicals (and their complexes), aromatic amines and their derivatives, emulsifiers and solvents of various nature.
2. The mechanisms of the studied processes have been established in the whole and as elementary stages, their basic kinetic characteristics have been determined.
3. Equations to describe the behavior of the studied chemical systems in the reactions of polymerization proceeding in various physicochemical conditions have been derived.
4. Scientific principles of regulating polymer synthesis processes have been elaborated to allow optimization of some industrial technologies and solving most important problems of environment protection.

REFERENCES

Culture and Ecology. *Searching for Ways of New Ethics Formation* / Ed. E.R. Melkumova. Moscow, 1996. [Russ].

Goldfein M.D., A.V. Ivanov, N.V. Kozhevnikov. *Fundamentals of General Ecology, Life Safety and Environment Protection*. New York. Nova Science Publishers, 2010. [Russ].

Goldfein M.D., N.V. Kozhevnikov, A.V. Ivanov, N.I. Kozhevnikova, A.N. Malikov, L.G. Timush. *Fundamentals of Ecology, Life Safety, and Economico-Legal Regulation of Nature Usage*. Moscow, 2006. [Russ].

Goldfein M.D., N.V. Kozhevnikov, N.I. Kozhevnikova, S.E. Molina, L.G. Timush. *Fundamentals of Ecology, Life Safety, and Environment Protection*. Saratov, 2000. [Russ].

Miller G.T. Jr. *Living in the Environment: Principles, Connections, and Solutions*, 14th Edition. Thomson Brooks/Cole, 2005.

Moiseyev N.N.. *Man and Biosphere*. Moscow, 1990. [Russ].

Nebel B., R.T. Wright. *Environmental Science: The Way the World Works*. Prentice Hall College Div, 1999.

Odum E.P.. *Ecology*. 5th ed. Brooks Cole, 2004.

Reymers N.F.. Hopes for Mankind Survival. *Conceptual Ecology*. Moscow, 1992. [Russ].

Ursul A.D., A.L. Romanovich. *Safety and Sustainable Development*. Moscow, 2001. [Russ].

Vernadsky V.I.. *Biosphere and Noosphere*. Moscow, 1989. [Russ].

Vernadsky V.I.. *Live Substance and Biosphere*. Moscow, 1994. [Russ].

Vernadsky V.I.. *Origin and Eternity of Life*. Moscow, 1989. [Russ].

Chapter 10

CONCEPTS OF GLOBAL STUDIES

10.1. GLOBAL PROBLEMS, SYSTEMS AND PROCESSES

Global problems become more and more aggravated and realized by mankind, and the global studies are now one of the bases of our new outlook. In this regard they become more and more close with such scientific multi-interdisciplinary leads as ecology, synergetics, and evolutionism. *Global problems, processes, and systems* are interrelated, but each notion requires resolving its own specifics of globality criteria. It is impossible to discern only natural or anthropogenous global systems, processes, and problems in modern natural sciences since they are of complex natural and anthropogenous character; further, the interaction and co-evolution of natural and anthropogenous components will progress. It is necessary to consider the synthetical character of scientific perception of globalization, meaning the need in a wide systematic and evolutionary approach, complex research conducting by representatives of the natural sciences and humanities. At the same time, modern approaches having gained development in other scientific leads should be more widely involved in global studies. These are, e.g., the synergetic, ecological and evolutionary approaches to cognition of global processes through the co-evolution concept, and the usage of mathematical methods of modeling of global systems and processes as well.

All global problems manifest themselves on both local and regional levels. This means that the most fundamental and practically effective studying of global systems, processes, and problems is achievable at mass collecting and careful analysis of the material from local districts and separate regions with its subsequent comparison, generalization, and comprehension on a global scale.

Problems of Modern Global Studies

Now, solving the morphological task prevails in global studies, which, along with positive moments of studying of global systems and processes of the modern world, can be considered as one of the principal problems. In more detail the problem of crises should be also considered with the use of the results of development in such leads as the theory of catastrophes and research of the influence of crises on the global Earth evolution. Global crises cannot be considered as merely political, financial, ecological, and psychological ones since the crisis phenomena have complex character. Of greatest practical interest is the problem of the development of modeling and prediction of global systems and processes. An effective way out from the global economic crisis including its humanitarian and world-outlook components can be found, finally, in the way of creation of a uniform synthetic globalistic concept. Such a concept is absent in modern natural sciences.

The origin and formation of global studies as an interdisciplinary sphere of scientific knowledge date back to the last third of the 20th century. Its emergence was caused by the process of integration of various sciences for solution of complex problems on a planetary scale. Scientific interpretation of the content of the term "*global studies*" dates to the 1990s when the scientists' basic attention was switched from studying specific global problems to establishment of the sense of the globalization phenomenon. Development of global studies was related to clarification of the increasing importance of the problems of nuclear safety, natural resource limitation, ecological balance, the need to overcome the backwardness of developing countries, etc. The objectivity of the interdisciplinary nature of global studies is caused by the following conceptual directions:

- critical analysis of the accruing threats to mankind from scientific and technical progress;
- futurological forecasts widely spread in scientific knowledge (the 1960s);
- design of the theory of dynamics of social systems which the methodological principles of global modeling could be developed within;
- the biosphere-to-noosphere transformation concept (Vladimir Vernadsky, the 1930s) which is actually a hierarchical system of philosophical, political, natural-scientific, technical and economic aspects of global problems of the present. These leads of scientific thought underlie modern global studies.

At the first stage (the late 1960s–late 1980s), global studies were generally concerned with comprehension of global problems rather than globalization processes. The second stage of their development coincided with the process of disintegration of the socialist system, which caused a new arrangement of power on the international scene at the end of the 20th century.

Just this served as a push to turn scientific thought from studying of the consequences to analysis of the causes and essence of the changes occurring on our planet. In the 2000s, a third stage of development of the global studies begins, associated with studying of globalization as the process of formation of all-planetary structures in various spheres of social life. Now it is considered that global problems grow out of the global processes making negative impact on social life, and their positive consequences represent themselves as global factors of the progressive development of mankind.

The question of the status of global studies is very important. Some scientists consider it as a scientific discipline, others—as a sphere of public practice. Third ones call global studies an overdisciplinary area of scientific knowledge.

Among the achievements of global studies are the development and formation of an interdisciplinary communication language, specification of basic notions and categories. With the amplification of scientific knowledge and ordering of the information on the occurring world events, new theoretical models of global tendencies, problems, processes, and systems are created.

The modern global studies develop in the following directions.

1) *Philosophico-methodological direction.* The philosophical bases, essence, and genesis of global processes are studied; most important sociopolitical and economic transformations are analyzed.
2) *Socio-natural direction.* It contains the problems of security with raw material, energy, water, soil, and other resources; representatives of natural, technical, social sciences, political and public figures are engaged in solution of these problems. Their works are directed on the development of principles and methods of optimal interaction between the society and nature, ecologization of productions and rational environmental management.
3) *Culturological direction.* It is connected with studying of the problems arising in the sphere of scientific and technical progress, population, health care, culture, law, education, and other areas of public life. When using the complex and system approaches, the *problem field of global studies* is formed, which includes: diversity of the development models of mankind; forms of the interaction of national states under the conditions of globalization; transformation of political, economic and socio-cultural processes; the essence of socio-natural interactions.

10.2. Some Aspects of Global Evolutionism

Three basic notions underlie evolutionary natural sciences, namely: change, movement, and development. *Change* characterizes the state of moving matter; *movement,* as one of the main forms of the existence of matter, represents any processes of its change; *development* is a certain type of movement characterizing the direction of change of the content of a material object, phenomenon or process.

Any development accompanied by transition from the lowest to the highest is progressive, otherwise it is regressive. The concept of development has integrative character and promotes both the process of synthesis of natural-scientific, humanitarian and technical knowledge and strengthening of the interrelation among the sciences about nature and the man, which leads to the emergence of new generalizing and often universal notions, concepts, and theories on development. The modern natural and humanitarian fields of knowledge are not mere adequate reflection of the properties of matter, the mechanisms of phenomena and processes, and the relevant laws but also their changes at different stages of development and evolution.

One of the central questions of universal evolutionism is about the degree in which the terrestrial and space aspects of evolution are combined in the problem of development. Their in-depth study by philosophy (which has generalized particular scientific and universal ideas of development) should be the main condition of such association.

The geocentric and space development processes are elements of a whole, namely, the *Universum,* under which the Universe as all real things available to observation is understood, and the Universum in a wide philosophical sense as well. Just the philosophical outlook and methodology have appeared that sound basis which has tied the terrestrial and space, particular and general development processes into a scientific concept of evolution.

The emergence of evolutionism as a scientific outlook is associated with the works of such philosophers as Kant, Herder, Leclerc, Buffon, Regel et al. (the 18–19th centuries AD). The term "evolution" in relation to the human body was first introduced by Hale (1677) and in a wider biological sense (development of the whole live matter) by Spencer (1852). In physical and chemical sciences the idea of development got accustomed hardly. Up to the second half of the 20th century dominated the initial abstraction of a closed reversible system, in which the factor of time does not play any special role. Principal changes occurred in this regard upon transition from classical physics to relativistic and quantum one. The notion of entropy and the idea of irreversible time-dependent processes were introduced in thermodynamics. Classical thermodynamics studied only closed equilibrium systems, while non-equilibrium processes were considered as merely perturbations, minor deviations which can be neglected. Penetration of the evolutionary idea into geology, biology, sociology, and other humanities at the end of the 19th century–first half of the 20th century occurred quite independently in each area of knowledge (though some interference of sciences took place). By the end of the 20th century, natural sciences approached the design of a *uniform model of universal evolution*, revealing of the general laws of nature to tie the origin of the Universe (cosmogenesis), the emergence of the Solar System (heliogenesis) and Earth (geogenesis), life emergence (abiogenesis and then biogenesis), and, at last, the man and society (anthroposociogenesis). Just such a model is developed within the framework of the concept of global evolutionism, in which the Universe appears as a developing-in-time natural whole, and all the history of the Universe from the *Big Bang* to the emergence of mankind is considered as a uniform process of evolution of space, chemical, geophysical, geographical, biological, and social systems continually and genetically connected among themselves.

Global-scale problems arose only in the second half of the 20th century when it was necessary to pass from the abstract-dialectic picture of development to a universal model which would proceed from the basic concepts of modern natural sciences, including interdisciplinary sciences. The development of evolutionary natural sciences in nature can be tracked since elementary particles. In comparison with them, atoms are more complex and organized systems. Such transition from one qualitative level to another one was one of the first steps known to the man in animate nature evolution. Next step of development of material systems are molecules whose quantity of types, as a result of the variety of chemical bonds, appears immeasurably more than that of atoms.

As a result of molecular evolution, even more complex organic compounds up to biopolymers (proteins, polysaccharides, polynucleic acids) have appeared which support main vital functions in biological systems. Along with natural selection, biochemical evolution (generally caused by changes in the mechanisms of biochemical reactions) is a most important motive power of animate nature evolution, which some tendencies are characteristic of. First, it is the trend of temporary acceleration confirmed by the hierarchy of the origin of live beings of various levels of organization (unicellular, multicellular, micro-organisms, plants, and animals, including the highest mammal primacies—anthropoids and the man). Second, during different historical periods the evolution of various populations and species occurred with different rates. The conditions, in which structures and structural elements of development steps can be in nature, are in a quite wide range. Elementary particles exist almost everywhere, except physical vacuum (they can be therein but cannot constitute it). Atoms exist within a shorter criteria range: only on planets, in cosmic space, and on the surface of some types of stars.

Molecules are there as well, but the range of their conditions is even narrower than for atoms. This results from the fact that expansion, for example, of a temperature range could cause dissociation of molecules into atoms (either ions or free radicals), which is caused by the character of the chemical bonds in molecules. Therefore, the concept of universal evolutionism indicates quite distinct orientation of the development of nature, i.e. an increased ordering of the structural organization. Besides, such factors play the determinant role in animate nature evolution as the emergence of new structural elements (as a result of mutation processes), information exchange among them (due to features of the genetic code), etc.

Information which can be stored, processed, transferred etc. is considered a universal criterion to unequivocally determine many characteristics of the development of material objects in nature. When studying material objects, phenomena, and processes, of primary interest is information storage, i.e. only that part of information which according to Brillouin's terminology is called as *fixed*. It may not be transferred from one object to another since the main thing consists in that it acts as a certain characteristic of a rather steady distribution or a variety of elements in the structure of object. It is meaningful to analyze only finite, limited in space in time systems and at a certain level because without these restrictions the quantity of information (or the quantity of variety) in any object, owing to the inexhaustibility of matter and its development and the general relation among all phenomena, is infinitely great. Owing to application of the information criterion, new possibilities to determine the degree, rates, and direction of evolution of material systems in nature emerge. Information accumulation manifests itself, first, at development on one level within which that object is above by development which contains more information; and, second, this law is also true for the principal line of progress connected with transition from one step of matter development to another, higher step.

The whole history of the Universe, from the singularity till the emergence of the man, appears as a uniform process of material evolution, matter self-organization and self-development. A qualitatively new level of matter organization will be finally self-affirmed only when it is capable of effectively making use of the whole previous experience of the historical development of matter. In this regard, of great importance is the so-called *anthropic principle* stating that the emergence of mankind as a cognizing subject was possible owing to that some key properties of our Universe were those which they actually are. This principle reflects the deep intimate unity of the regularity of the historical evolution of the Universe, the Universum, and preconditions of the emergence and evolution of the organic world up to anthroposociogenesis.

According to this principle, there is some type of the universal system links determining the integral nature of the existence and development of the Universe as a certain systematically-organized fragment of the infinitely diverse material nature. Global evolutionism, on the one hand, provides a notion of the world as of some integrity, allows considering the general laws of life in their unity and, on the other hand, focuses modern natural sciences on detection of specific regularities of matter evolution at all levels and stages of its self-organization.

All this points to modern searching for both general philosophical laws and development tendencies, and universal and invariant directions and regularities of evolution of the whole material world, which can predict further interaction of mankind with the environment of Earth and space.

10.3. Biosphere as a Global Ecosystem

Living organisms are a function of the biosphere and are closely, materially and energetically, related to it.

Vladimir I. Vernadsky

The notion "biosphere" was introduced by Suess (1831–1914) in 1875. In his understanding, the biosphere is a set of all living organisms, the live shell of Earth. Another viewpoint proceeding from interrelation of living and inanimate organisms was put forward as swell. Humboldt (1769–1859) developed his idea of the integrity of nature, interrelation of all its processes and phenomena. Dokuchaev (1846–1903) in his doctrine of landscape and geographical zones developed his concept of the unity of all elements of the terrestrial surface, which had the most fundamental justification and development in Vernadsky's (1863–1945) doctrine of the biosphere. The biosphere arose about 4 billion years ago. Most generally, the organization of the biosphere is unity of the biotic (living) and abiotic (lifeless) elements involved in the sphere of life.

The *biosphere* (from the Greek *bios* – life, *sphaera* – a ball, sphere) is the shell of Earth covering the atmosphere to the bottom border of the ozone layer (20–25 km), the top part of the lithosphere, and the whole hydrosphere. Its composition, structure, and energetics are determined by the cumulative activity of living organisms. The basic source of the biogeochemical activity of organisms and, therefore, the power basis of the biosphere is solar energy.

The modern biosphere represents a geologically instant cut of the long-existing and developing panbiosphere, i.e. sets of the former biospheres (the term was coined by Boris Sokolov in 1981). For the biosphere development, the following tendencies (by Edward Kolchinsky, 1988) are characteristic.

1. Gradual increase in the total biomass and bioproductivity.
2. Progressive accumulation of solar energy in the surface shells of Earth.
3. Increase in the information capacity of the biosphere which manifests itself in growth of the variety of organic forms, in increase in the number of geochemical barriers, and in strengthening of the differential nature of the physico-geographical structure of the biosphere.
4. Amplification of some biogeochemical functions of live substance and the emergence of new functions therein.
5. Intensification of the transforming influence of life upon the atmosphere, hydrosphere, and lithosphere; increase in the role of live substance and the products of its activity in geological, geochemical, and physico-geographical processes.
6. Expansion of the sphere of action of biotic circulation and complication of its structure.
7. As all evolving systems are not immortal, have the "beginning" and "end" of their existence, both "ascending" and "descending" trends in the biosphere development are possible.

Except the listed tendencies, Nikolai F. Reymers (1994) has formulated the following principles of the biosphere development.

Eco-historical principle. Geological processes and phenomena constantly changed in time, so no processes of the past can be completely identified with the modern ones (especially in a quantitative sense).

The principle of global closure of biogeochemical circulation. During the biosphere evolution, the fraction of the biological component in closure of the biogeochemical circulation of substances regularly increases.

Ervin Bauer's principle of self-development of biological systems (1935). Development of biological systems is a result of their increased external work, expressed in their impact upon the environment.

The principle of a catastrophic push. Any natural or natural-anthropogenous catastrophe always leads to essential reorganizations of the biosphere to adapt it to these new conditions of the environment.

The principle of continuity and discrete development of the biosphere. The process of slow evolutionary changing is regularly interrupted by rapid development and fast extinction phases.

The evolution of the biosphere can be considered from multi-stage positions. The stages of the biosphere development somewhat correspond to the geochronological divisions, i.e. with the stages of development of our planet as a whole. In the course of the activity of organisms, basic changes in the abiogenic components of the biosphere occurred. E.g., oxygen appeared in the atmosphere; the "ozone screen" emerged to protect living objects from short-wave UV radiation; a part of carbon dioxide was preserved as deposits of coal and other minerals. Some substances were removed from active biotic circulation (limestone, ores, saltpeter, etc.). The biosphere stretches up to approx. 35 km from the terrestrial surface (to the top border of the ozone screen); the bottom one is limited by organic deposits on the oceanic bottom. The biosphere and its live objects are in indissoluble interrelation. On the one hand, the vital activity of developing species determines features of the biosphere; on the other hand, just the structure of the biosphere causes the possibility of survival and development of separate species and populations. The existence of a set of living organisms is a specific feature of the biosphere. Few tens of millions animal species and plants live on Earth; more than 90% of them inhabit land. The basis of the Earth biomass is green plants capable of photosynthesis. The organic world of our planet exists as biotic circulation to provide continuity of the processes in the biosphere. Especially high in biotic circulation is the role of those microorganisms who promote decomposition of organic substances into mineral salts and elementary organic compounds which then are utilized by green plants for subsequent synthesis. The power source of biotic circulation is sunlight, but only about 0.2% of solar energy goes to synthesis of organic substances, and its most part is spent for abiotic processes.

The development of the biosphere is associated with the contradiction between the boundless ability to reproduction peculiar to living organisms and the limitation of its natural and resource potential within a specific existential range. This contradiction is resolved by the use of new substances, power sources, and information. Hereditary variability acts as a necessary precondition of development, and natural selection is as a mechanism to fix new information. Living organisms cannot exist in the biosphere only on the basis of internal cycles of self-control; they develop under the influence of constantly changing, in space and time, external conditions, and their evolution is a function of not only the biosphere but also adaptation changes of heredity. Living beings possess information accumulated in the course of evolution by previous generations. The organism processes this information, stores, and again transfers outside, already taking into account certain responses to the corresponding external influences; and heredity serves as the keeper of information, expressing the organism's evolutionary experience to provide both the internal integrity of the organism and its harmony of relationships with the biosphere. The character of the environment can influence strong impact upon any organism. If the corresponding influence of the biosphere falls outside the durability limits of some living organism, this leads to breakage of its integrity and irreversible changes. The organism not only selectively reacts to the environment requirements but also can influence it.

For the biosphere as a whole, an exclusive stability is characteristic and only the appearance of the man made preconditions to breakage of the historically developed balance of the biosphere and its subsystems.

Now there are a number of concepts of the biosphere.

The *biogeochemical concept* was developed by Vladimir Vernadsky in the 1920s. It is underlain by revealing of interrelation of space, geochemical, and geological evolution factors. According to this concept, the biosphere includes the following main components:

- *live substance* is a set of all living organisms;
- *biogenic substance* is created and processed by organisms (coal, bitumen, lime, etc.);
- *abiogenic substance* is formed without any participation of living organisms (tectonic activity products, space bodies, etc.);
- *bioinert substance* is a result of the joint activity of organisms and abiogenic processes (water, soil, etc.).

The basic functions performed by live substance in the biosphere are the gaseous (all organisms), oxygen (chlorophyll plants), oxidizing (bacteria, mainly autotrophic), calcic (algae, bacteria, etc.), regenerative (bacteria), concentrating (animals and plants), destructive (bacteria and fungi), reductive decomposition (bacteria) ones, metabolism and breath processes (all organisms).

The high organization of the biosphere is created and preserved by the activity of live substance whose main form is its biogeochemical work, i.e. implementation of irreversible and unclosed circulations of substance and transformation of energy flows among the structural components: rocks, natural waters, gases, soils, vegetation, animals, and micro-organisms. Studying of biogeochemical cycles as unclosed circulations helps one to get into the essence of the processes of organization of the biospheric shell more deeply. Every subsequent state of the biosphere repeats no previous one. Involvement of some material and energy flows into migratory cycles and release of other ones from biogeochemical cycles lead to continuous renewal of the biosphere, promotes its progressive evolutionary development and complication of live substance. Concentrating space energy and transforming it to the active energy of terrestrial processes, living organisms aspire to the maximum manifestation of this effective energy in the processes of exchange, circulation, and biogeochemical cycles. Just the direct links and feedback of such cyclicity make the mechanism of functioning of the biosphere and the essence of its organization. The development of the biosphere is associated with its complication, which is expressed in the increase in the variety of live beings, the increase of their organization and expansion of the life spread sphere. Therefore, the main idea of the biogeochemical concept of the biosphere is recognition of the evolution of species in organic unity with geochemical and geological factors, and their interrelation manifests itself in biogenic migration of atoms in the biosphere.

The *thermodynamic concept* (Gladyshev) is caused by biospheric processes obeying the second law of thermodynamics. For a living object, entropy growth, i.e. disorganization (chaos), is accompanied by its destruction and leads, finally, to full degradation (death). The live substance of the biosphere makes essential impact not only upon geochemical migration on the terrestrial surface but also upon energy transformations occurring in the biosphere. The presence of live substance gives special features (not characteristic for other spheres: the hydrosphere, atmosphere, etc. of our planet) to the energetical phenomena in the biosphere. Just within this concept the interrelation between the second law of thermodynamics and the phenomena occurring in the biosphere is established, and the biosphere has thus reached a high level of its organization. The form of implementation of the laws of thermodynamics in the biosphere essentially differs from their manifestation in physical systems within inanimate nature. Living systems not only obey the thermodynamics laws but also adapt for them, and also "conserve" free energy capable to further transformations. The second law of thermodynamics is known to be fully applicable to isolated and closed systems only. And as the local ecosystems united in the biosphere are open onto space, the biosphere has the basic possibility to realize a tendency to entropy reduction, i.e. to ordering growth. Therefore, the biosphere as a whole is capable of self-renewal, and the difference between the live and the lifeless consists just in the ability of the living to create order from chaos and to transform entropy to anti-entropy (negentropy). Besides, the thermodynamic concept of the biosphere causes the property of adaptation of organisms as a natural anti-entropic state of any living system.

According to Schmalhausen's (1884–1963) *biocybernetic concept*, the full cycle of transformations in an elementary cycle of the evolutionary process forms four principal structural levels of life, namely: 1) hereditary information transfer through a zygote and cellular divisions; 2) information transformation in individual development (phenotype realization); 3) return information transfer through phenotypes of the individuals constituting the population-specific level of life organization; 4) transformation of return information in the biogeocenosis and hereditary information record in DNA molecules. The mechanism of life self-control and self-development is explained on the basis of these structural levels. The biocybernetic concept quite fully reflects the functional nature of self-organization of living systems which is possible owing to their open character, i.e. exchange of not only substance but also energy and information with the environment. Biological effects of the interaction of living systems with the environment depend not so much on the value of their energy as on the information received by them, according to which energy is organized with the aim of its economy. Such is the main criterion of self-organization in the biosphere. Therefore, the hierarchical system of the biosphere, in which both functional subordination of the parts to the whole and their relative autonomy are combined, has appeared a very constructive model of self-control and self-management of living systems.

Computer and mathematical modeling of biospheric processes. One of the first works in this direction was done by Volterra (1860–1940) who offered a "predator–victim" mathematical model with the use of a system of differential levels. In the 1960–80s, computer modelings of biospheric processes became more active and were developed within the framework of global modeling. Lyapunov (1911–1973) developed his ideas of cybernetic modeling of biological communities, from elementary ecosystems to the biosphere as a whole. Computer models of the biosphere were designed by Moiseyev (1917–2000). One of the difficulties in modeling of biospheric processes consists in the plurality of parameters necessary to be considered to design an adequate model.

10.4. Concept of the Noosphere

At the initial stage of transition of the world community to a new form (strategy) of development it is important to join together the scientific developments already existing in various countries and scientific communities and their results, which would promote reconstruction of an image of that future which received the name of the sphere of reason, or noosphere (the term coined by Vernadsky). In this sense it is possible to define sustainable development as a future model of the socio-natural process whose purpose is to preserve mankind and the biosphere and to create the noosphere.

The *noosphere* is a hypothetical future state of the society and its interaction with nature, where reason will occupy a priority place. The idea of the "kingdom of reason" was stated as early as during the Age of Enlightenment, when global environmental problems were not thought of yet. The Enlightenment thinkers gave the world an ideology penetrated by the belief in human reason, urged to provide progress of the society and to release it from the fetters of ignorance, myths, superstitions, ridiculous traditions, and tyranny. However, the "Enlightenment reason" appeared unclaimed, and the spontaneous procession of capitalism along our planet has led, in fact, to the formation of industrial and consumer values instead of the noospheric ones. Concepts of the sphere of reason, or the noosphere, offered by its founders essentially differed among themselves. However, all of them had some general content, namely, the belief in the power of reason (not only human one) and representation of reason as the main driving mechanism of the subsequent evolution. Though the term "noosphere" was used by Constantine Tsiolkovsky, nevertheless, he also spoke about the role of reason in the Universe, believing that the thought would become not only a geological factor but also a determinant in space evolution. In the course of real progress on the way to the noosphere, science will play a dominating role, as well as the conjugated intellectual-spiritual and moral-cultural forms of activity. In the future noosphere, social and information factors will appear more important and development-determining ones in comparison with material-energy components, forming the sphere of reason by noospherogenesis. Noospherogenesis, arising in separate places, countries, and regions of our planet, finds a global dimension and becomes a planetary process.

Creation of the sphere of reason should actively influence the globalization processes by rationalizing them with the moral and fair principle and by promoting harmonization of the relations in the "man—society—nature" system. Now the viewpoint is more and more affirmed that preservation of the biosphere and ensuring its highest possible natural evolution is necessary for optimum noospherogenesis as a planetary process. Just the biosphere is that basis of the existence of all life on our planet and further development of reason, if the latter will create mechanisms of an essential decrease of the anthropogenous pressure upon the environment.

Vladimir Vernadsky put forward an idea of formation of the noosphere and social autotrophy as the main direction of the further development of mankind. His followers have essentially developed his ideas and shown that the era of formation of the noosphere and transition of the world community to sustainable development are related processes. The idea of survival and continuous (steady) development of the civilization as co-evolution (mutual preservation and co-development, mutual adaptation) of the man and the biosphere means, at the same time, mankind's advance to the sphere of reason where rational management of the society and nature interaction will be provided. In the course of real progress on the way to the noosphere, science will (as Vladimir Vernadsky assumed) play a priority and dominating role to form the sphere of reason.

Preservation of the biosphere and survival of mankind determine the basic principles of the noosphere formation (noospherogenesis), which generally reduce to the following.

First, it is the *principle of co-evolution of the society and nature*, i.e. their mutual preservation and development, when mankind can exist infinitely long on our planet, and the biosphere, owing to the future essential decrease in the anthropogenous pressure, will keep its stability and evolve under its own laws. Second, it is the *principle of equal opportunities* in satisfaction of the vital requirements (including those in environmental conditions and natural resources) for both present and future generations. Third, it is the *justice principle* in its broadest sense, which quite often in the documents of the United Nations is considered the basic one for sustainable development.

Justice until recently was thought as compliance between the actions of people and their public recognition. However, social expansion of the principle of justice requires its further (already socio-natural) interpretation. Namely, some features of justice should be transferred on all live beings whom the man must not use with impunity in his egoistic purposes and even must not kill. Fourth, it is the *priority of moral and fair reason and intellectual spiritual needs and values over material ones*. Sustainable development only then becomes noospheric development (noospherogenesis) when it is spiritualized by moral reasons of eco-humanistic orientation and just it appears a priority means of the subsequent continuous development of our civilization in harmony with the environment.

It should be noted that the notions of spirit, spirituality differ from those of reason, mind, and intelligence since the former ones focus no attention on rational comprehension of the world. Spirituality is opposite to every material and natural rather than to irrationality. These differences of the discussed notions lead to a conclusion that in the course of noospherogenesis not simply the rational principle but just the rational-moral one will dominate.

One could expect that noospherogenesis will be associated with the emergence of a new form of rationalism, namely, noospheric rationalism, and also with new forms of rationality to express the role of reason in increasing the efficiency of social activity. The emergence of new versions of rationalism is supposed to be connected with these or those forms of rationality. E.g., new principles of rationality already appear in technocratic rationalism, somewhat distinct from the previous ones.

The critical period in the modern science and culture essentially loosens the bases of both the former versions of rationalism and the corresponding forms of rationality. Therefore, modern rationalism, considering time requirements more and more, appears very clearly connected with survival of our civilization and solution of global environmental problems. This in full measure applies to the expected, more adaptive forms of rationalism connected with the formation of the noosphere.

10.5. Concept of Co-Evolution

Under *co-evolution* (co-development, conjugated development) is understood the development of two or more systems in their interaction causing development of the interaction itself (its forms and mechanisms) in time. The main achievement of co-evolution processes is harmonization of the relationship among the systems. Co-evolutional processes come to light at most various levels of the organization of matter and are now an object of attention of many fields of knowledge. The notion of co-evolution, having arisen in the life sciences ("co-evolution of symbionts", "predator and victim co-evolution", etc.) has now intense spreading.

The concept of co-evolution indicatively works, e.g., at a level of the ecology of organisms, to explain the development of the "predator–victim" system, i.e. continuous improvement of both components of this system. In the "parasite—host" system, natural selection should theoretically promote survival of less virulent (not dangerous to the host) parasites and more resistant (steady against parasites) hosts. The parasite gradually becomes a *commensal*, i.e. safe for the host, and then they could become the *mutuals*, i.e. such organisms who promote mutual prosperity, such as fungi and photosynthesizing bacteria together forming lichens. But so occurs not always. Parasites are an inevitable, obligatory part of every ecosystem. The co-evolutional "race of arms" promotes a wide variety of ecosystems. Parasites hinder their hosts' destruction of other species. The concept of co-evolution explains altruism facts in animals: their care of children, aggression elimination by demonstration of "pacifying poses", obedience to the leader, mutual aid in difficult situations, etc.

Co-evolution is a most evolutionally favorable version of the relationship among systems in the "coexistence—interaction—co-evolution" triad. The strongest interactions are of the same sign, namely: competitive (–;–) and symbiotic (+; +). All possible combinations of signs of positive, negative, and neutral systems' impact upon each other (+;–; 0) exist in nature. The most progressive result is brought by symbiosis (+; +), i.e. mutually beneficial cooperation. In the evolutionary plan, this version of relationships is obviously most favorable over all others and is capable of entailing, as a result of co-evolution, the emergence and development of new forms of life on both organismic and cellular levels (symbiogenesis).

At the all-planetary level of organization it is possible to speak of the co-evolution of the shells of our planet (the geospheres). Each geosphere spatially covers a certain part of the planet and, in the general modeling view, has the outlines of a shell (sphere). The geospheres are basic elements of Earth and together determine a structurally functional unity of our planet as a most complex system. During the co-evolution of the geospheres, a regularity of co-evolutional resonance manifestation acts: for realization of co-evolutional possibilities of two systems, coincidence in time of certain parameters of both systems is necessary. E.g., it is supposed that in the early Precambrian, various versions of the prokaryotic forms of life were appearing consistently in time, but, in those conditions of the environment, had gained no development. The realization was successful only when two parameters coincided in time, namely: the emergence of next version of eukaryotic systems in the biosphere and the achievement of the oxygen concentration in the atmosphere necessary for providing profitability of the eukaryotic mechanism, i.e. the Pasteur point (which, in turn, is also a consequence of the biosphere development). A similar situation is supposed concerning the origin of multicellular organisms as well.

10.6. THE "MAN–BIOSPHERE" SYSTEM AND GLOBALIZATION PROCESSES

Functioning of the biosphere is provided with that the live substance is involved into biotic circulation, without breaking the basic principles of its organization. And only in the course of the emergence and development of the man, the historically developed natural cycles gradually start to be broken, thereby undermining the biological bases of the existence of life.

In the course of "Neolithic revolution", i.e. transition from the appropriating economy to the producing one, when the ancient man not only collected nature's gifts but also carried on primitive agriculture and cattle breeding, first centers of ecological tension are observed. The emergence of fire and first forms of activity allowed making the man's role in nature active. Calculations of some researchers show that the activity of the primitive man could lead to the disappearance of a number of species of large mammals (mammoths, etc.); the corresponding negative ecological processes were of local character.

In the course of the formation and development of the modern forms of civilization, since the industrial revolution of the 18th century, and especially by the beginning of the 20th century, the man becomes the force of "planetary scale" (Vladimir Vernadsky). The production-economic and socio-cultural activity of mankind has more and more negative impact upon the biosphere. By the beginning of the 1970s, a representation at the level of the world community (the United Nations Conference on the Human Environment, Stockholm, June 1972) was formulated, according to which mankind was approaching a dangerous line in its relationship with the biosphere: the increase in the carbon dioxide concentration in the atmosphere and the tendency of increasing average planetary temperature; the growth of "ozone gaps"; the increases emissions of human activity products into the biosphere; the fast rates of disappearance of traditional types of flora and fauna; deterioration of the conditions of the human life quality.

Analysis of world social and ecological processes has allowed to draw a conclusion that preservation of the current trends in the world development will lead to exhaustion of natural resources, pollution of the biosphere and, finally, to ecological collapse at the end of the 21st century ("The Limits to Growth" report of the Club of Rome, 1972). However, the global character of the environmental problem has regional features as well. In the developed countries, a more and more considerable part of their economic potential (above 2% of GNP) goes for ecological purposes. As a result, despite development contradictions of the market relations, the "man—biosphere" system in the majority of the developed countries is gradually stabilized. On the contrary, in the developing countries their weak economy and high demographic growth create no positive conditions for overcoming of the biospheric tension sharpness. Solving the environmental protection problems is not easy in the conditions of the modern Russian Federation. Our economic difficulties do not allow allocating the necessary amounts of nature protection capital investments (less than 1% of GNP is allocated for these purposes). As a result, the ecological situation has tense character in many regions of the Russian Federation.

The "man—biosphere" problem has global character, and environmental problems, affecting, to a greater or lesser extent, the majority of the countries of the world community, require realization of the principles of combination of national, regional, and international programs of the rational usage of natural resources and nature conservation. Globalization as a natural-historical spontaneous process began since mankind had acquired an aspiration to become a uniform civilization. As a phenomenon, globalization has systematic character, since one of its main results is mankind's finding its unity. However, this should finally appear the socio-natural unity realizing the purpose of survival of our civilization, solution of global problems, and eliminations of threats and dangers to the civilization existence. Globalization from these positions is represented as some new stage in the processes of change of the civilizational forms, stages, and periods of development, which would be considered as development of the uniform mankind. Just collision of these or those tendencies of the civilization with planetary limits generates global problems, which would be possible to consider as the end of the "extensive" and "pre-global" history of our civilization.

Globalization should not be considered as process only. At the same time, this is a regularity of human history because it expresses an objectively real, stable, necessary, and essential relation in the "man—society—nature" system, expressed in this system's finding its global socio-natural integrity. It is expedient to expand our understanding of the globalization process, having given a socio-natural dimension to this notion. It is necessary to make such system expansion on the essence of research of the problem of "globalization", especially considering prospects of its expansion through sustainable development. If the transition to sustainable development appears a solution of the global problems, then the globalization turn towards the purposes of this development will also promote survival of our civilization and the biosphere preservation. To a certain extent, globalization through sustainable development can be treated as some kind of "advancing" globalization, different from the "backward" globalization in the model of unstable development. As owing to transition to sustainable development it is necessary to consider the possibilities of satisfaction of future generations' requirements and interests in the increasing measure, the turn of the globalization vector towards the future is obvious. Just for this reason, globalization through sustainable development will appear the purposeful socio-natural futurized activity of the whole world community.

In the modern social-scientific literature there are a number of theoretical constructions, and among them there are concepts of post-industrial society, information society, society of risk, ecological society, the noosphere, etc. A characteristic feature of all these conceptual formations is that they are compatible to the processes of globalization and every concept applies for seeing the future in the borders of its subject field, and more often similar vision is connected with this or that tendency of globalization. At the same time, not each of these directions corresponds to the model and strategy of sustainable development since they were designed within the previous model. Therefore, the historical and research choice appears related with their attitude to the mentioned by standard-target model of future socio-natural development.

There is a viewpoint that the post-industrial society and the society with sustainable development are actually the same. Meanwhile, just the post-industrial society is the finishing stage of the model of unstable development under which a society with the non-coevolutional production way of interaction with nature is understood. The post-industrial society at a certain stage of its development could become a precondition and a launch pad to transition to the "steady society" if it starts to realize the principles of the latter one, which have global character. All countries of the world community should pass to sustainable development irrespective of what stage of development they are at: agrarian, industrial or post-industrial.

The information society serves a "bridge" from the post-industrial society to sustainable development, which, at certain transformations, can pass test for the criteria of sustainable development. The assumption that the information society will continue its development in the further globalization process is considered rather reasonable. The information society first develops spontaneously, within the natural-historical process of the formation of a post-industrial society; but at its certain stage, the necessary global preconditions and conditions will be created connected with novel information and telecommunication technologies and the corresponding infrastructure owing to which there appears a planetary and space cyberspace allowing to make decisions and at a global level.

10.7. Problems of the Urbosphere in Global Studies

The global socio-natural process of urbanization inevitably results in the need of cooperation among urban structures (exchange of material resources, population migration, mechanisms of interaction of intellectual character). The interaction between urbanized territories is carried out, first of all, by means of the creation and development of specialized international organizations. A special place among them is occupied by the United Nations Educational, Scientific and Cultural Organization (UNESCO). The evolution of the urbosphere as a global system generates very sophisticated and historically multilayered urban agglomerations. In this regard, even within structurally-functionally complete urbosystems there often emerge problems requiring special research and decisions. This is exemplified by the contradiction between preservation of culture monuments and their usage, and also with the possibility of harmonious coexistence of elements of the past and the present within the format of urban space. In UNESCO's official documents it is stated that the development of research and administrative activity to solve global problems of cities is carried out directedly, on the way "to the city of the 21st century". Along with the already performed particular actions to solve the social, cultural and environmental problems of urban life, UNESCO considers it necessary to support the measures offered by the Second United Nations Conference on Human Settlements (Habitat II), by expansion of research, personnel training, public information, and implementation of pilot projects in the sphere of management and environmental protection in cities. The complexity and multi-aspect nature of the global urban problematics in the activity of international organizations are obvious. As the former Secretary-General of the United Nations Boutros Boutros-Ghali noted at one of the top-level Conferences on urban problems, "*there are hard questions to answer. How can we improve the governance and finance of human settlements? What policies are needed to improve conditions for the poorest people, families and communities? How can we ensure basic hygienic conditions in urban areas, while avoiding long-term damage to the environment? Can we ensure that, by a target date, adequate shelter will exist for all? What must be done to mitigate the effects of natural disasters and war? Can the cycle of deprivation, conflict, devastation and failure to develop be broken?*"

The common nature of the global problems of cities within the limits of the urbosphere as a whole makes quite logical the foundation of various urban associations. Such alliances are possible on both world scale ("The World Heritage Cities Programme") and regional one. It is possible either under the aegis of UNESCO (as "Southern Italy's world heritage sites"), or on a different basis (as "Association of Cities of the Volga Region" in the Russian Federation). A special sphere is UNESCO's scientific and educational activity carried out by organizing a number of thematic network projects and structures. Urbanization can be considered as a global socio-natural process (urban population growth, increase of the role of cities and their territorial development, spreading of the urban way of life and urban culture) connected with concentration of activity in relatively few centers of social and economic development.

Urbanization is characterized by several development stages. The first stage (the 19th century) outlined main contours of the geographical and demographic trends in the cities of Europe and North America. At the second stage (the first half of the 20th century), urbanization was spread almost on all regions of the world, accompanied by the trends of growth rate acceleration of the urban population. The third stage (the second half of the 20th century) is characterized by an ascending dynamics of the process of urbanization, a considerable growth of the urban population, and the emergence of new cities. At complex analysis of the problems of global studies and the urbosphere it is possible to resolve two main research directions. The first one assumes studying of *global cities* as urban settlements extending their influence (economic, political, etc.) in the all-planetary scale. The second direction considers the urbosphere as a global system with its structuring and functioning performed through networks of urban settlements. Now, above half of the Earth inhabitants live in towns and cities, while the urbanized territories occupy only about 3% of the terrestrial land. But dangerous natural and technogenic objects concentrate just on these sites; a wide variety of social and ethnic groups, migratory streams, economic structures, political trends is concentrated. Such a situation shows the multifactor tension and increased risks in cities, which is connected with manifestation of the global problems having most various character. The variety of types of urban settlements built by the human society since its emergence and interconnected by elements of the engineering infrastructure (transport and information networks, elevated and underground communications) is a unique property of any urbanized space. Urban settlements not simply fill space but are interconnected and develop by some uniform regularities, which allows one to speak of the formation of an all-planetary system of cities and towns. Consideration of the space saturated with urban structures as a special shell of our planet also seems logical. The modern shell structure of Earth is a result of its global evolution. New, natural-anthropogenous shells are formed at interaction of natural (the geologo-geomorphological substratum, soils, microclimate, etc.) and anthropogenous factors; one of them is the urbosphere. Urban spaces are, in the long term, capable of absorbing elements of the anthroposphere (technosphere) and represent growth points of the urbosphere. The urbosphere is spatially overlapped with the geographical shells, covering the top part of the lithosphere and the lower part of the atmosphere. No exact borders of the urbosphere, their geometrical features and changes of these borders and of the general configuration of the urbosphere are established by now.

The main feature of the urbosphere consists in that it interacts and co-evolves with all the shells of our planet. Therefore, the phenomenon of the urbosphere should be considered with the use of the multi-interdisciplinary approach only. Providing of sustainable development of the human civilization is also associated with urbosystems. The regularities of the urbosphere agree with the concept of co-evolution of the geospheres and with the fractal-cellular model of the geospheres. *Evolution of the urbosphere* (urbospherogenesis) characterized by a number of trends (increase in the sizes and volume in time, complication of the top and bottom borders) is a global socio-natural process. Development of the "settlements—city—

urban agglomeration—megapolis—megalopolis" system is accompanied by evolution of the forms of the urbosphere from the dot ones to the planetary and lens-looking ones. This means that the modern cities are preceded by a number of point urban settlements, subsequently united into a uniform urbosystem. The capacity of the urbosphere changes depending on features of development of the urban agglomeration. The global regularity of the urbosphere development is shown in changes in the sizes, outlines, volume, and substance and communication redistribution both inside the urbosystems and within the boundary geospheres.

It is also expedient to include an *antiurbosphere* into the global socio-natural system. In the Earth sciences and global ecology, studying of crisis and catastrophic phenomena in the geographical man-coped space is topical. First of all, it concerns large settlements, namely, urban systems where geoecologically dangerous natural and natural-technogenic processes develop. The man-used underground and elevated space of urban systems is considered as one of the shells of our planet arisen and developing in the formed noosphere's contact with the other geospheres and the noosphere. At the same time, the political, economic, social and ecological factors of development of urban systems are capable of creating conditions for the emergence of elements of antiurbospheric conditions therein (destructions and degradation of the urbosphere) and of leading to death of cities, finally. The top border of the content of the antiurbosphere (the visible zone of dispersion) is almost absent, and the bottom border is expressed by elements of neglected technical communications and networks. A *dead* city exemplifies the antiurbosphere. In a broader sense, the dead city is a product of the global process of urbanization, in which the processes of functioning of elements of the urban ecosystem (the natural or technogenic media) are considerably broken or absent; more specifically, it is a settlement with a completely destroyed infrastructure (left by its inhabitants owing to the influence of political, historical, economic, ecological or other factors). The causes and mechanisms of development of neglected settlements is one of the most sophisticated questions in the modern global ecology connected with the subjects of study in the Earth sciences, ecology, humanitarian disciplines, and some interdisciplinary scientific leads (historical urboecology, urbogeomorphology, engineering geoecology). The scenario of the future and features of social and economic development of *dead* cities are ambiguous. Studying of the problems connected with the functional usage of the territories of dead urbosystems is one of the strategic problems of any state. An optimum solution of this problem depends on development and introduction of a complex of priority actions, which should essentially change the urbofunctional condition on the territory of cities, determine the necessary directions to develop the industrial, transport and social infrastructure, provide the urban population with ecologically pure water and food, etc. The greatest effect in studying of global processes in large cities can be reached by means of two interconnected scientific and methodical approaches. The first one is based on detailed studying of the factual material directly over the territories of particular large cities with subsequent comparison and historical analysis. The second one assumes analysis of published and archive sources on urban problematics with analysis of the historical dynamics of parameters of global urban systems. An effective technique of assessment of features of the development of urbosystems is detailed analysis of the life comfort of the population in cities with detailed zoning of urban areas as well. A city in which all its components (architectural, industrial, social, intellectual, economic, etc.) are harmonious and develop co-evolutionally can be considered as a steadily developing system. Special attention should be given to forecasting of negative irreversible processes in the development of urbosystems and their networks.

REFERENCES

Abylgaziev I.I., A.V. Ivanov, I.V. Ilin, Yu.N. Sayamov, A.D. Ursul, A.S. Sheshnev, I.A. Yashkov. *Global Urban Systems*. Moscow, 2012. [Russ].

Abylgaziev I.I., R.R. *Gabdullin, I.V. Ilin, A.V. Ivanov, I.A. Yashkov. Global* Socio-Natural Processes and Systems. Moscow, 2011. [Russ].

Barlybaev Kh.A., *General Theory of Globalization and Sustainable Development*. Moscow, 2003. [Russ].

Borinskaya S.A., *Evolutional Principles in Nature and Society*. Moscow, 2002. [Russ].

Fedotov A.B.. *Global Studies*. Principles of the Science of the Modern World. Moscow, 2002. [Russ].

Ilin I.V., A.V. Ivanov. *Introduction into Global Ecology*. Moscow, 2009. [Russ].

Moiseyev N.N.. Universum. Information. Society. Moscow, 2001. [Russ].

Chapter 11

Synergetics As a New Interdisciplinary Lead

Synergetics can be considered as a forum, on which scientists of different disciplines have met to exchange their ideas of how to treat big systems.

Hermann Haken

11.1. Notions and Definitions

The notion "synergetics" was offered by Haken in 1973 to denote the scientific lead establishing the most general laws of self-organization, i.e. the phenomenon of coordinated action of the elements of a complex system without being governed from the outside. *Synergetics* (from the Greek *synergetikos* — joint, coordinated, cooperating) is the scientific lead studying links between the elements of a structure which are formed in open (biological, physicochemical, geologo-geographical, etc.) systems owing to intense exchange of substance, energy, and information with the environment under non-equilibrium conditions. In such a system, the coordinated behavior of its subsystems is observed, as a result, the orderliness degree increases (the entropy decreases), i.e. some self-organization process develops. Synergetics is underlain by the thermodynamics of non-equilibrium processes, the theory of stochastic processes, and the theory of nonlinear oscillations and waves. Synergetics can be considered as an interdisciplinary scientific lead used to explain the emergence of macroscopic phenomena as a result of interactions among microscopic elements. The synergetic model of self-organization is now the most generalizing model to describe mutual order—chaos transitions in evolution processes of complex (both natural and social) systems. Therefore, the modern synergetics is a system of concepts describing regularities of the structure, functioning, and evolution of open systems.

The coexistence of systems is always caused by the existence of links, which are divided into two types. *Direct link* is such one when an element (A) acts upon another one (B) without any response. The action of a wood circle on a grassy plant casually grown under its cover and the impact of the Sun on terrestrial processes exemplify such links. At *feedback*, the element B answers the action by the element A. Feedback can be positive or negative. The positive feedback entails the process amplification in one direction. Bogging of some territory (e.g. after deforestation) is an example of its action. Removal of the wood cover and soil consolidation usually cause water accumulation on its surface. This, in turn, gives water-accumulating plants the opportunity to grow, e.g., bog-moss whose water content by 25–30 times exceeds the net mass of their body (without water). The process starts to go in one direction: increase in moistening – depletion with oxygen – delay of vegetative remains decomposition – turf accumulation – further strengthening of bogging. The negative feedback acts in such a manner that in reply to strengthening of the element *A*'s action, the force of the element *B*'s action opposite by direction increases. Such feedback allows the system to remain in a state of stable dynamic equilibrium. This is the most widespread and important type of links in natural systems. The stability of ecosystems is primarily based on them. The predators and victims relationship is an example of such feedback. An increase in the numbers of victims as a nutrition resource creates conditions for reproduction and increasing of the numbers of predators. They, in turn, start to eat victims more intensely and to reduce their numbers. In the whole, the numbers of predators and victims synchronously oscillate within certain limits. Another example concerns the consequences of an increase in the carbon dioxide concentration in nature. Phenomena of local increases in the atmospheric carbon dioxide (e.g. owing to volcanic activization) took place in the history of Earth. It was followed by an increase in the intensity of photosynthesis and carbonic acid binding into organic substance, and its more intense absorption by the ocean. One more example is periodic increases of the levels of ground waters, which leads to their enhanced contact to the root systems of plants, increased evaporation through vegetation (transpiration), and the return of the ground waters to their initial level. Besides, an important property of complex systems is their emergence (see section 9.2) — a system is more than the sum of its parts. For example: few separately growing trees cannot make a "wood system" since they create neither a certain environment (soil, hydrological, meteorological, etc.) nor the wood-peculiar interrelations of various links to cause a new quality. Underestimation of emergence could lead to great failures at human intervention into the life of ecosystems or when designing systems to perform definite purposes.

Three concepts underlie the synergetic approach to consideration of systems, namely: non-equilibration, openness, and nonlinearity. *Non-equilibration* is such a state of a system at which its macroscopic parameters change, e.g., the composition, structure, or behavior. *Openness* is the ability of a system to constantly exchange substance, energy, and information with the environment and to possess both dispersion sources and dispersion zones. *Nonlinearity* is the property of a system to stay in various stationary states corresponding to various acceptable behavior laws of this system.

Synergetics considers three types of structures (patterns): *free*, *forced*, and *autopatterns*. *Free* structures are exemplified by whirlwinds behind a rotating screw, vortical rings at the outflow from a nozzle, habitual smoke rings. *Forced* patterns are convective cells created by a circular border (e.g., silicone oil containing aluminum powder is warmed up from below; the container is covered with a cooled glass plate to exclude surface tension and the formation of Benard's six-sided convection cells). Under *autopatterns* are understood localized spatial formations steadily existing in dissipative non-equilibrium media and independent (in finite limits) of any boundary and initial conditions. Autopatterns, in turn, are divided into static ones (with no movement therein), stationary (with continuous movements; e.g. rotation without changes in their internal structure), and dynamic (regular or chaotic changes in time therein).

Attractors are active steady centers of potential ways of the system's evolution, capable of "drawing" and organizing the elements of the system and its environment. They can be considered by the researcher as the purposes of an evolving system influencing the direction of the evolutionary process. The theory of attractors allows understanding the possibilities of control over complex systems. Attractors divide the space of all possible states of a system into attraction areas, having gotten into which the system inevitably evolves to the corresponding attractor. This causes a limit of any external impact on the system. Any influence can be effective and change the system's development tendencies, if only it draws the system to the area of influence of another attractor.

11.2. Self-Organization Concept

The processes of self-organization are considered in synergetics as key ones in the development of complex systems. The systems themselves and the processes, which they participate in, possess the following features.

1) Self-organization is the process of evolution of a system from disorder to order. Naturally, the entropy of the system in which self-organization occurs should decrease. However, by no means this contradicts the law of entropy increasing in closed systems, i.e. the second law of thermodynamics. Self-organization processes occur in open systems. If self-organization occurs in a closed system, it is always possible to outline an open subsystem, in which self-organization proceeds; at the same time, disorder increases in the closed system as a whole.
2) Self-organization proceeds in such systems whose state is essentially distinct from equilibrium at present. Equilibrium disturbance is caused by external influences. In the example with Benard's convection cells, the external influence is shown as heating of the vessel, which results in distinction of the temperatures in separate macroscopic areas of the liquid. In electric generators, an external influence (voltage created by a source) leads to a non-equilibrium distribution of electrons. The state of a system far from equilibrium is unstable with respect to a state near equilibrium, and, owing to this instability, processes leading to structure formation occur.
3) Self-organization is only possible in systems consisting of a large number of particles. In some cases it is obvious; e.g., macroscopic spatial structures contain a plenty of atoms and molecules. However, if we consider the self-oscillations of populations, it is possible to state that such self-oscillations are impossible at small number of individuals in the population. The matter is that the emergence of fluctuations (macroscopic inhomogeneities) is only possible in systems with many particles.
4) Self-organization is always related to spontaneous symmetry reduction. A beautiful symmetric snowflake has, nevertheless, lower symmetry than unstructured water vapor. The ideas of such changes in symmetry have gained good development in the modern theory of the microworld, and in the description of phase transitions in physics (e.g., the liquid–crystal transition). In general, the processes of self-organization are in many respects similar to phase transitions, so they are often called as kinetic phase transitions. The difference is in that microstructures (e.g., a crystal lattice) emerge upon phase transitions while the system remains homogeneous in a macroscopic volume.

5) The environment surrounding the system should be considered as a set of its constituent objects being in dynamics. The interactions of the studied objects in the environment can be either short-range or long-range.
6) The processes of organization and self-organization are distinguished. They have a common feature, namely, the increase in ordering owing to proceeding of the processes opposite to establishment of thermodynamic equilibrium of independently interacting elements of the environment (and moving away from chaos by other criteria). Organization, unlike self-organization, can be characterized, e.g., by the formation of homogeneous stable statistical structures.
7) Self-organization results in the emergence, interaction, interassistance (cooperation, co-evolution), and even regeneration of some dynamic objects (subsystems) which are more complex, in information sense, than those environmental elements which they arise from. The system and its components are essentially dynamic entities.
8) The directness of self-organization processes is caused by internal properties of the objects (subsystems) in their individual and collective manifestation, and influences from the environment which the system is "immersed" into.
9) The behavior of the elements (subsystems) and of the system as a whole is essentially characterized by spontaneity, i.e. behavioral acts are not strictly determined.
10) Self-organization processes occur in the environment along with other processes, in particular, those having an opposite direction; and separate phases of the system's existence can either prevail over the latter ones (progress) or concede them (regress). The system as a whole could have a steady tendency or undergo fluctuations to progressive evolution or to degradation and disintegration.

11.3. Chaos—Order Ratio

The phenomena of randomness and orderliness are extremely important in synergetics. Any nonlinear system during its existence period undergoes reorganizations in its structure, passing through chaos amplification stages. Both chaos and order are formed owing to self-organization processes. Chaos is a culmination state in fight of the "old" order with a "new" one, spontaneously aspiring to replace the "old". Synergetics considers several versions of chaos and order.

1. "*Chaos–order–God*" ("the central order" according to Werner Heisenberg) is a result of the mutually caused action of the general laws of matter development.
2. *Absolute chaos–order* characterizes vacuum, the absence of matter, time, and space. In vacuum, there are no material particles and, therefore, forces of interactions. If there exist the Universe borders which such vacuum begins behind, there are no events, and, therefore, no evolutionary processes therein. There are no laws of matter movement and structure formation in it.
3. *Thermodynamic chaos–order*. Some directed, irreversibly proceeding ordered process leads to it. Such chaos is formed owing to the irreversibility of processes and is consequently determined. The state of thermodynamic chaos could be quickly established in artificially created conditions closed for external influences, with no

supply of substance and energy to the system, so there is no energy in the system to produce work against irreversibility. The state of thermodynamic equilibrium can be considered as chaos because the location of any part of such a system is undetermined in both time and space. But it is caused by ordered movement of substance and energy as a result of the action of the laws of thermodynamics. The thermodynamic chaos–order can arise at ordered actions of high-rank systems only. Any technogenic or natural thermodynamic system if considered as closed can serve an example: a liquid overflowing in communicating vessels or natural reservoirs with establishment of a common level; the formation of plains on the terrestrial surface owing to substance redistribution from higher to lower levels, etc. In every case, in the absence of forces to interfere the development of irreversible processes, static thermodynamic equilibrium must be established at which any substance and energy redistribution in the system ends.

4. *Dynamic chaos–order* is formed as a result of ordered supply of substance and energy in quantities considerably exceeding those necessary for equilibrium. Another type of dynamic chaos–order is also possible, namely, the stochastic or nidal one. Dynamic chaos arises when substance and energy are brought to a system within a short period of time in such quantities which lead to breakage of the irreversibility of the course of processes and to destruction of the system. It is exemplified by rising of some blocks of the crust which occurs so promptly under the influence of endogenic forces that the exogenic processes have no time to process the substance and energy flow, and a poorly ordered conglomeration of geological bodies is formed. Catastrophic chaos is a special version of dynamic chaos. It is formed as a result of short-term supply of substance and energy into the system from its environment in quantities sufficient for complete destruction of the earlier existed ordered structures.

11.4. A Short Characteristic of the Theory of Catastrophes

The evolution of systems becomes complicated by crisis periods, catastrophic events, or sharp reorganizations. Such nonlinear phenomena are also considered by synergetics. The usual methods of control over a system, at which results are proportional to efforts, fail in crisis situations, and the manager needs to develop specific "nonlinear intuition" based on synergetic conclusions seeming sometimes paradoxical. Let us give some elementary qualitative conclusions from the mathematical theory of reorganizations (the theory of catastrophes) in application to a modeling nonlinear system being in an established steady state. Such a state is considered "bad" since there exists a better, preferable steady state of the system within visibility limits. The following features of development of the situation are possible under certain conditions.

1. Gradual movement towards the better state immediately leads to deterioration. The rate of deterioration at uniform movement to the better state increases.
2. In the process of movement from the worse state to the better one, the resistance of the system to changes in its state enhances.

3. The maximum of resistance is reached earlier than the worst state through which it is necessary to pass to achieve the better state. After passing of the resistance maximum, the state continues worsening.
4. At approaching the worst state on the reorganization ways the resistance, since some instant, starts to decrease, and as soon as the worst state is passed, not only the resistance completely disappears but the system starts to be attracted to the better state.
5. The degree of deterioration necessary for transition to the better state is comparable with the final improvement and increases in the process of improvement of the system. A poorly developed system can pass to the better state almost without preliminary deterioration while a developed system, owing to its stability, is incapable of such gradual, continuous improvement.
6. If a system has succeeded in transferring itself from its bad steady state rather close to a good one at once, jumpingly rather than continuously, it will further evolve towards its good state itself.

One must consider these objective laws of functioning of nonlinear systems. Theory also provides quantitative models, but the above qualitative conclusions seem more important and, at the same time, more reliable: they weakly depend on details of functioning of the system whose stability and numerical parameters could be poorly known. The mathematical theory of reorganizations is that part of the modern calculus without which no conscious control over complex and poorly known nonlinear systems is almost possible. No special mathematical theory is, however, required to realize that neglecting of the laws of nature and society (whether it is the law of gravitation, the law of value or the need in feedback, poor competence of the experts and lack of individual responsibility for the decisions made) will lead, sooner or later, to a catastrophe.

REFERENCES

Arnold V.I.. *Theory of Catastrophes*. Moscow, 2004. [Russ].

Chernavsky D.S.. *Synergetics and Information*. Moscow, 2002. [Russ].

Eigen M.. Self-Organization of Matter and the Evolution of Biological Macromolecule. *Die Naturwissenschaften*, Springer-Verlag, Berlin, Heidelberg, New York, 1971.

Galimov E.M.. *Phenomenon of Life: Between Equilibrium and Nonlinearity*. Moscow, 2001. [Russ].

Haken H.: *Synergetik*. Springer-Verlag, Berlin, Heidelberg, New York, 1982.

Kapitsa S.P., S.P. Kurdymov, G.G. Malinsky. *Synergetics: Forecasts of the Future*. Moscow, 1997. [Russ].

Kniazeva E.N., S.P. Kurdymov. *Laws of Evolution and Self-Organization*. Moscow, 1994. [Russ].

Kniazeva E.N., S.P. Kurdymov. *Synergetics: Time Nonlinearity and Co-Evolution Landscapes*. Moscow, 2007. [Russ].

Prigogine, I., I. Stengers. *Order out of Chaos: Man's New Dialogue with Nature*. Flamingo, 1984.

APPENDICES

APPENDIX 1. STRUCTURAL-LEVEL RELATIONS AND INTERACTIONS DETERMINING FORMATION OF MODERN CONCEPTS OF NATURAL SCIENCES

Natural sciences in the dynamics of the modern civilization

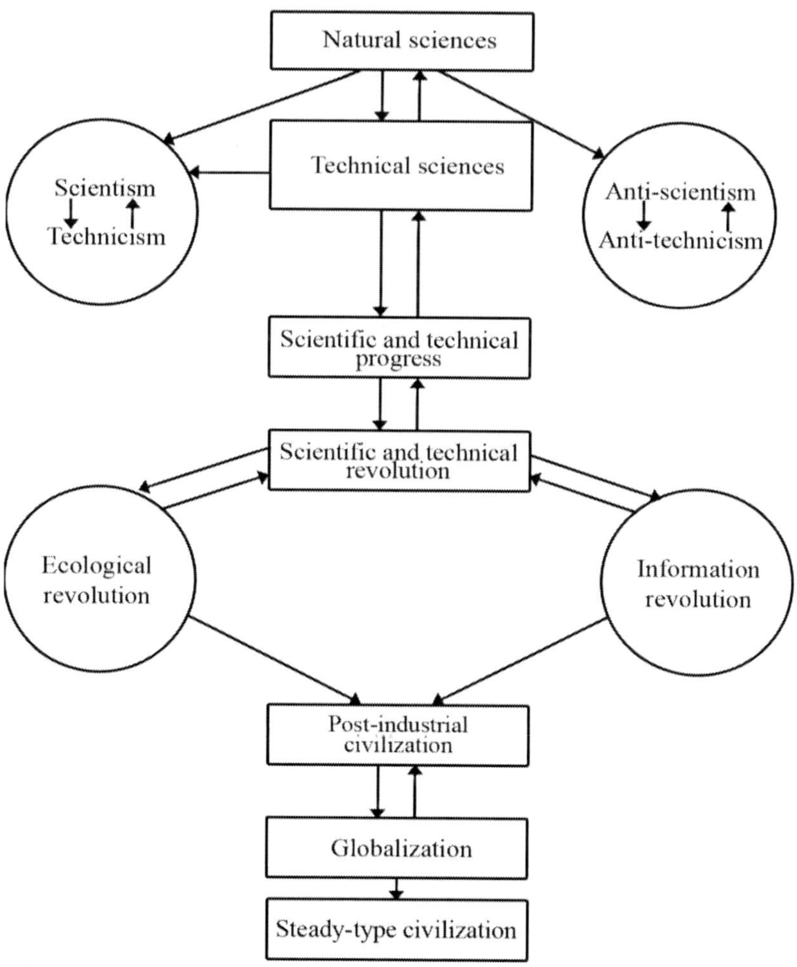

Historical features of natural sciences

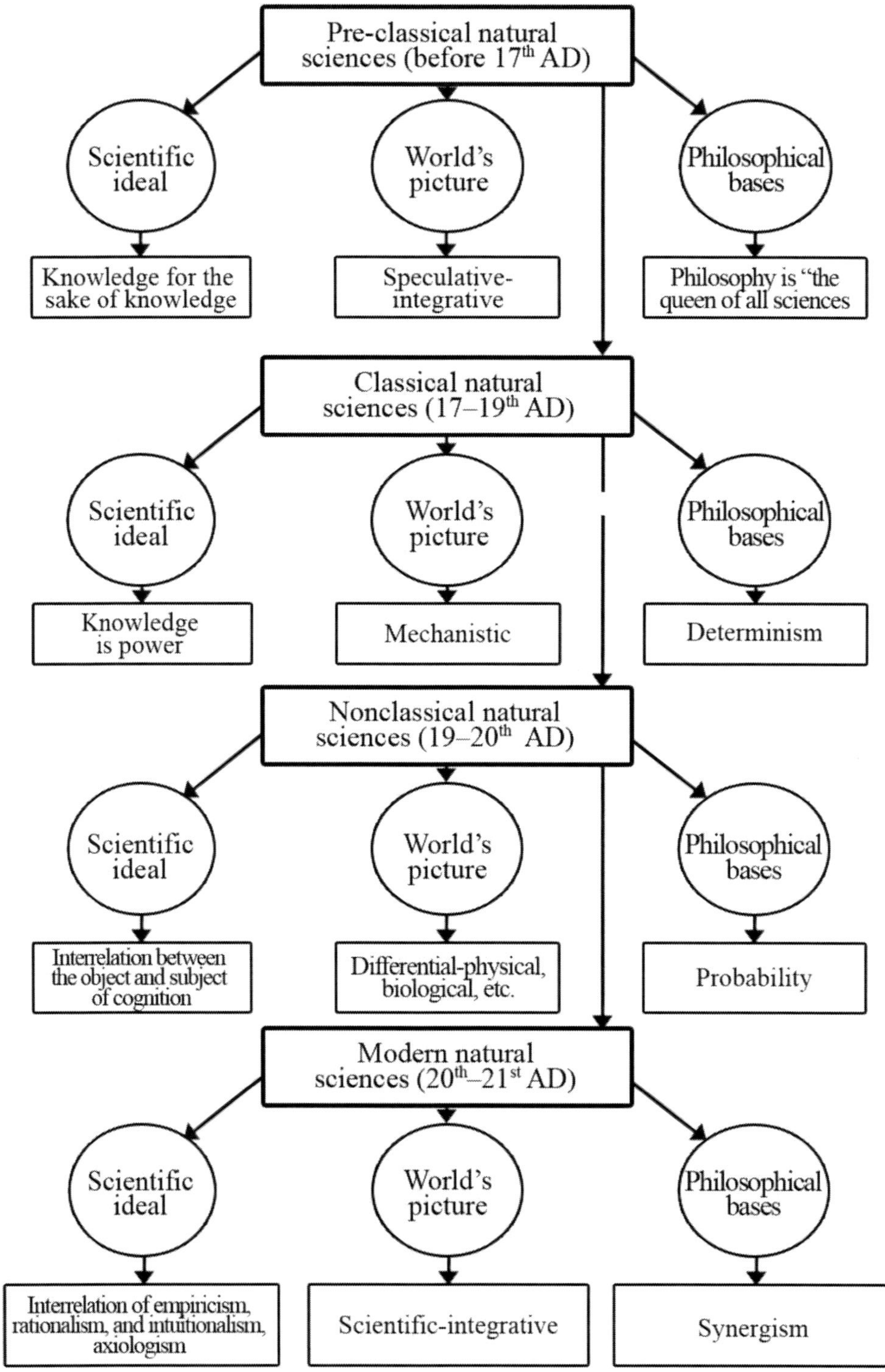

Empirical level of natural-scientific cognition

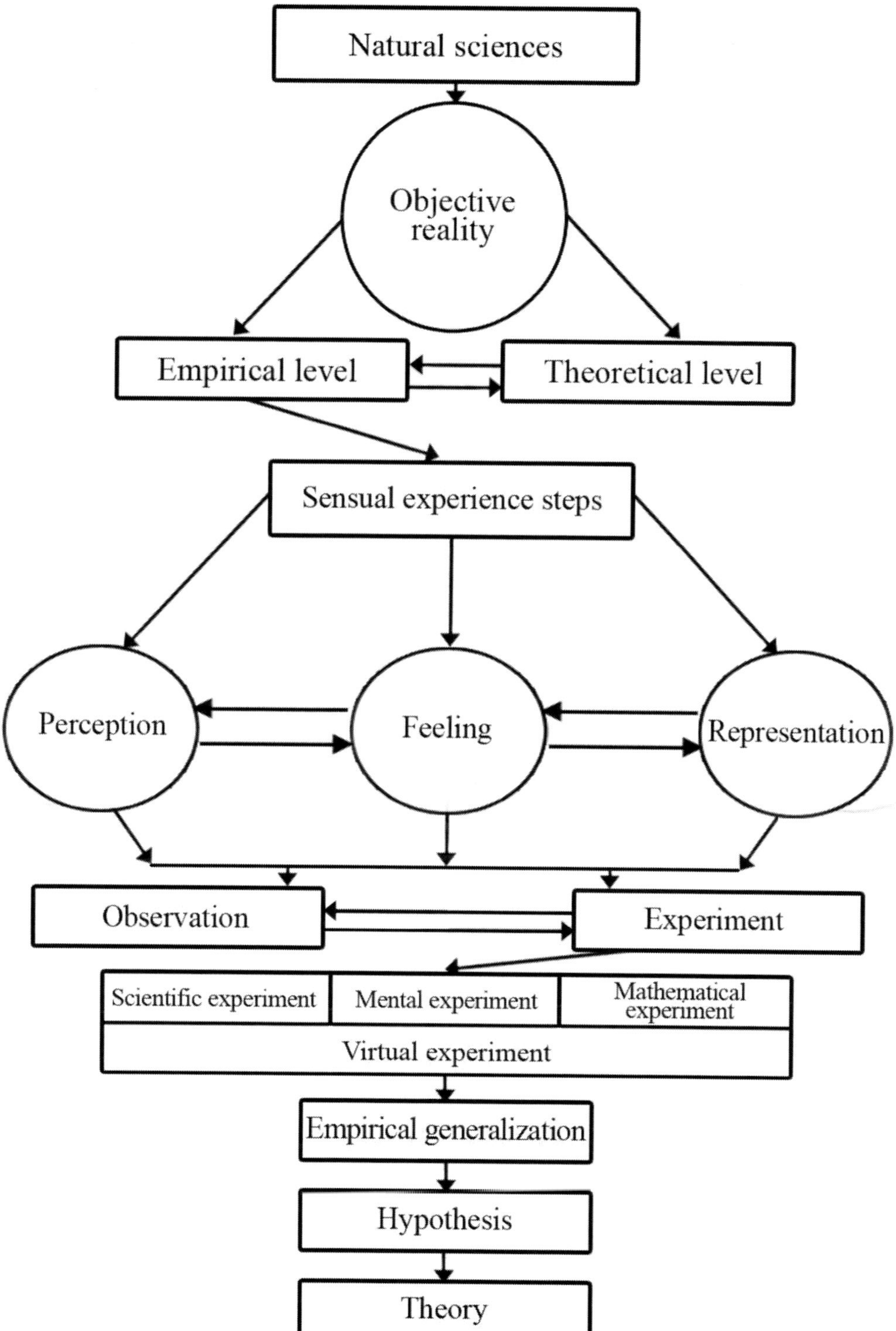

Theoretical level of natural-scientific cognition

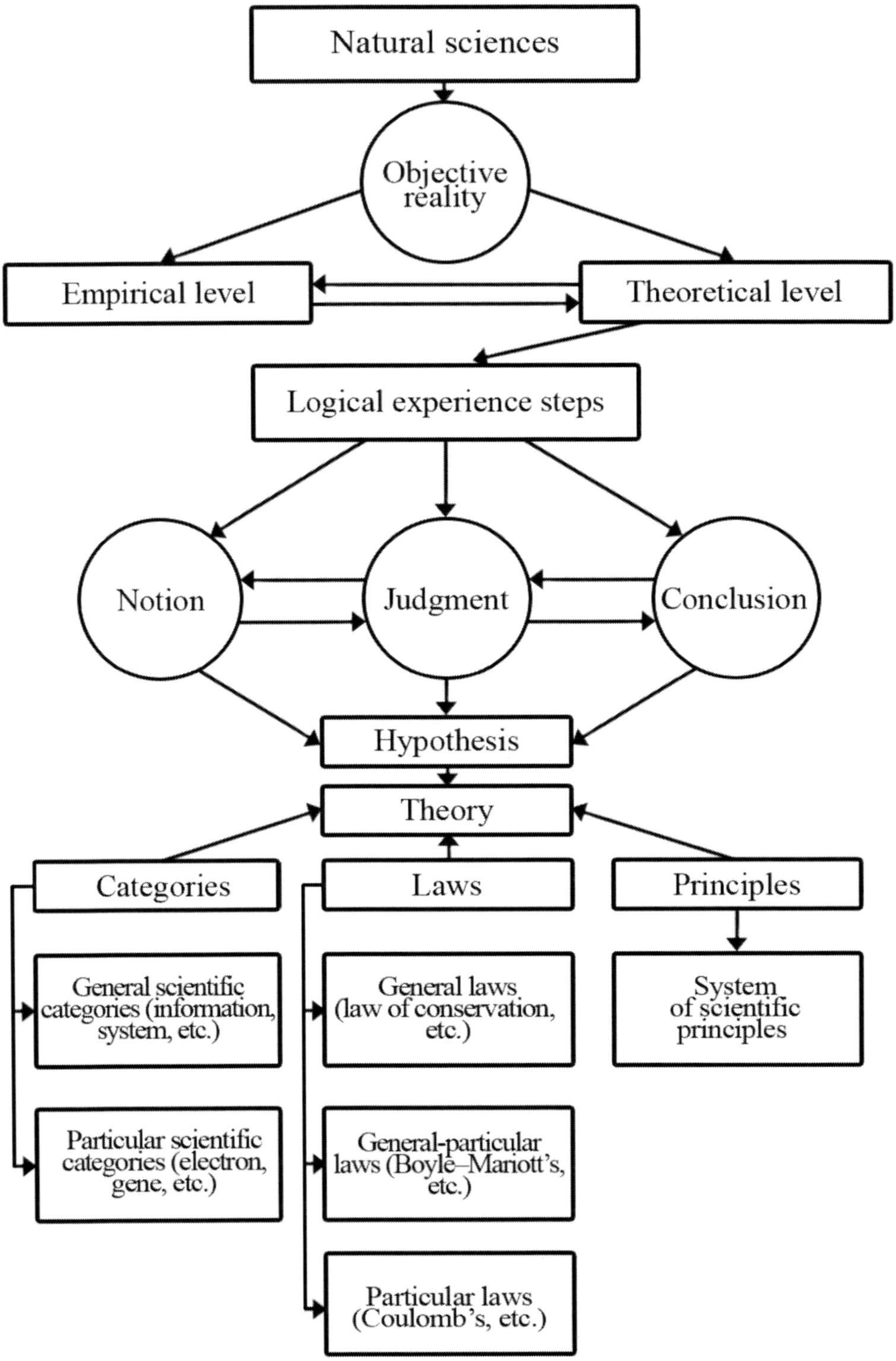

Design principles of a natural-scientific theory

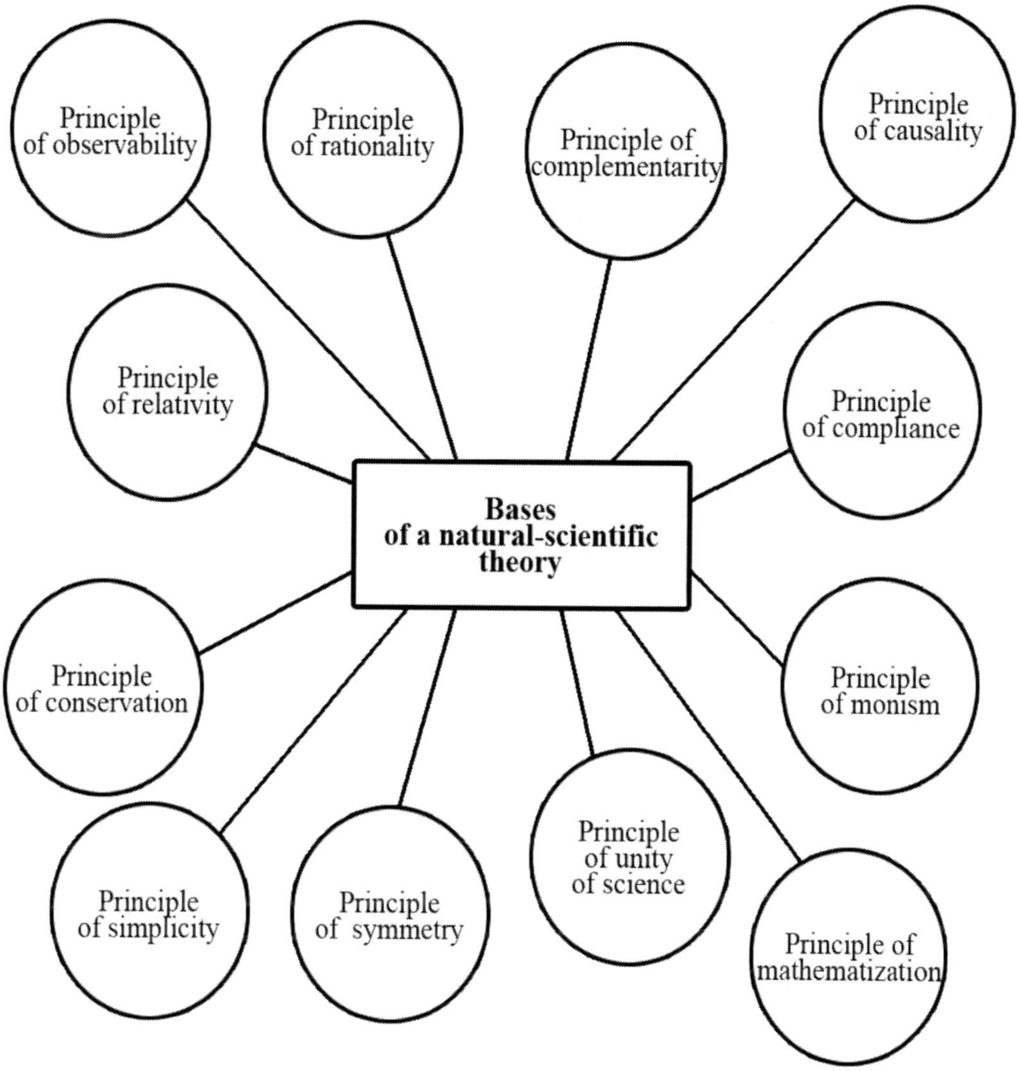

Basic development principles of the physical picture of the world

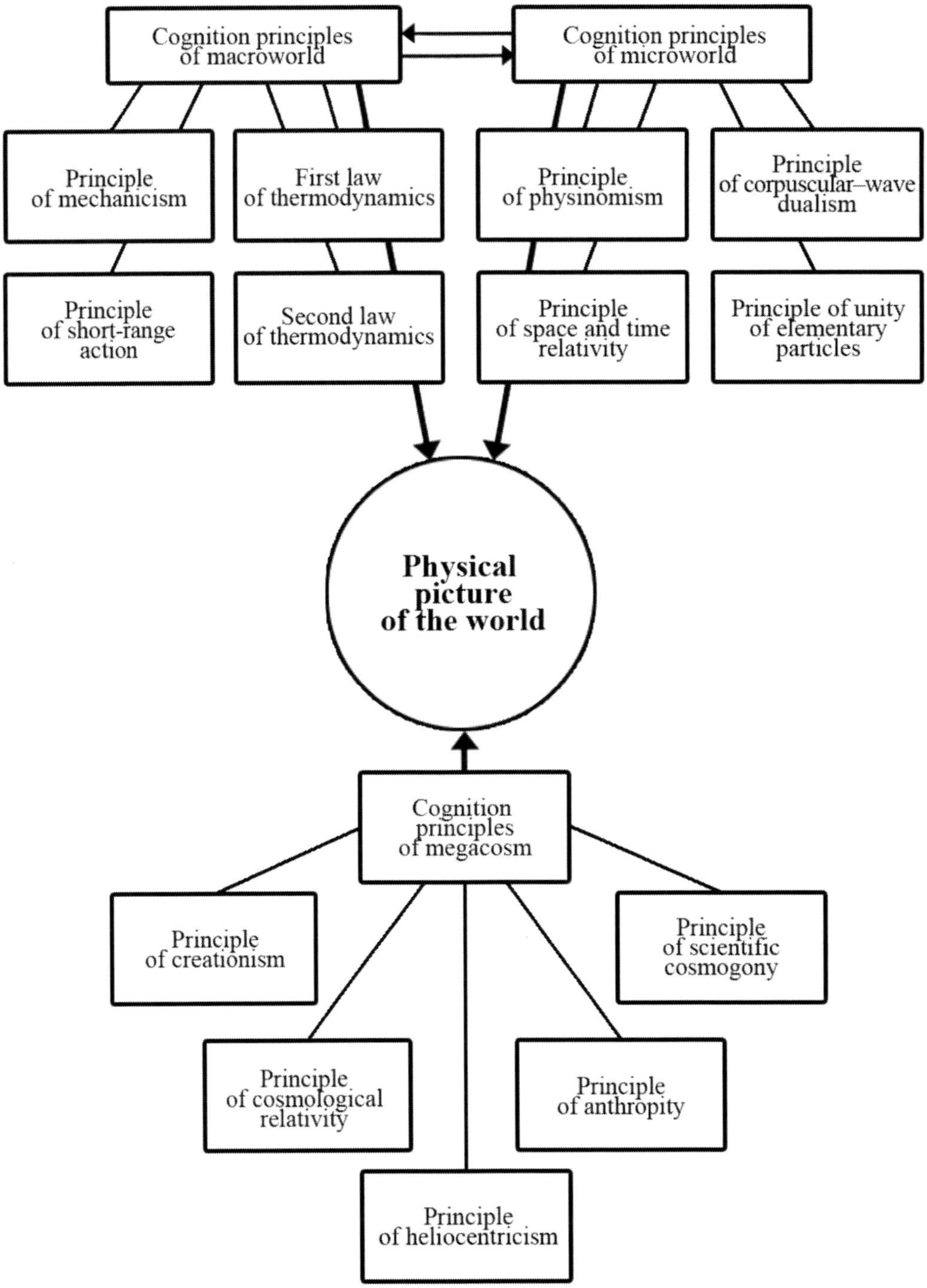

Dynamics of the chemical knowledge principles

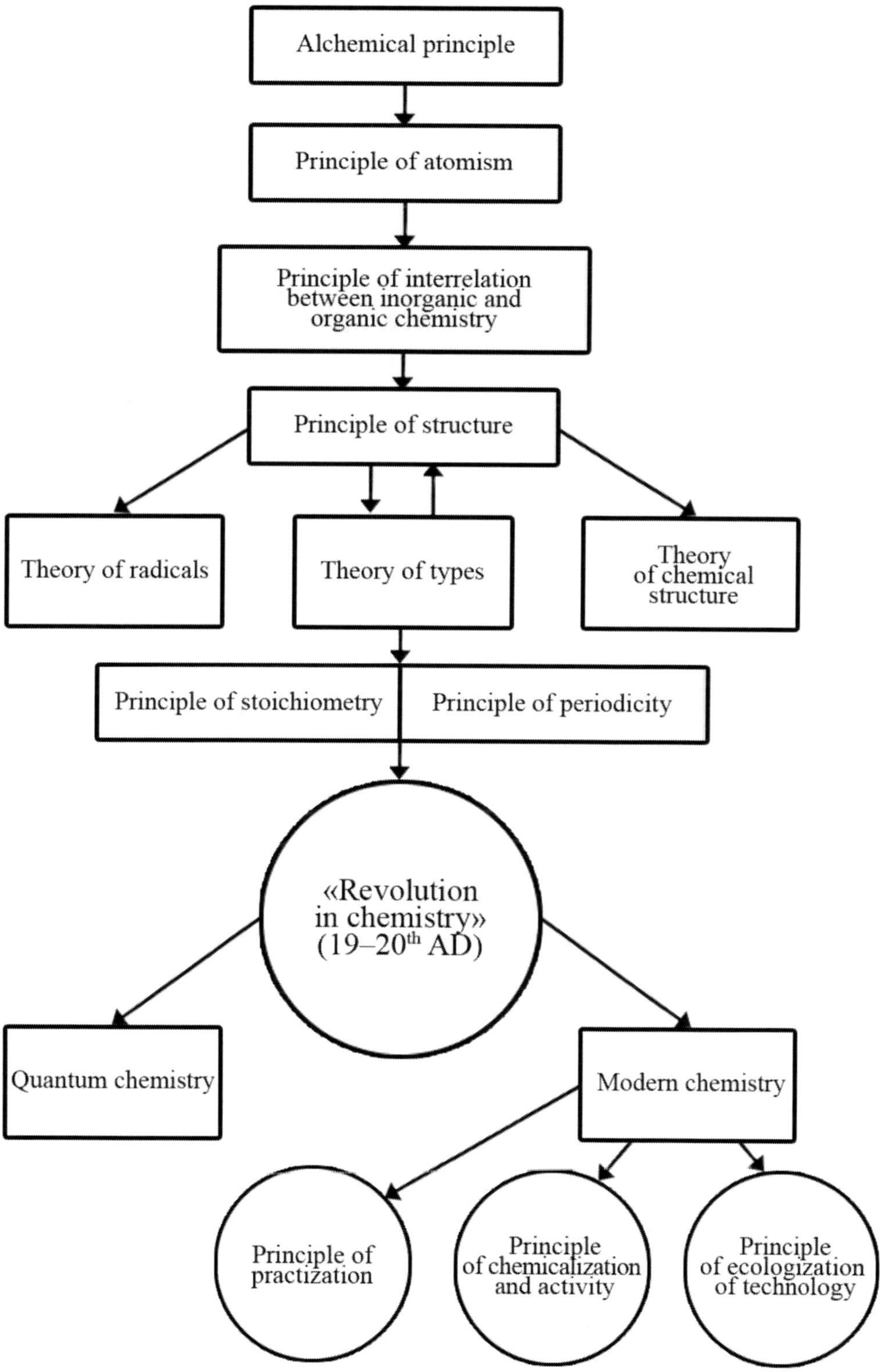

Basic concepts of life origin

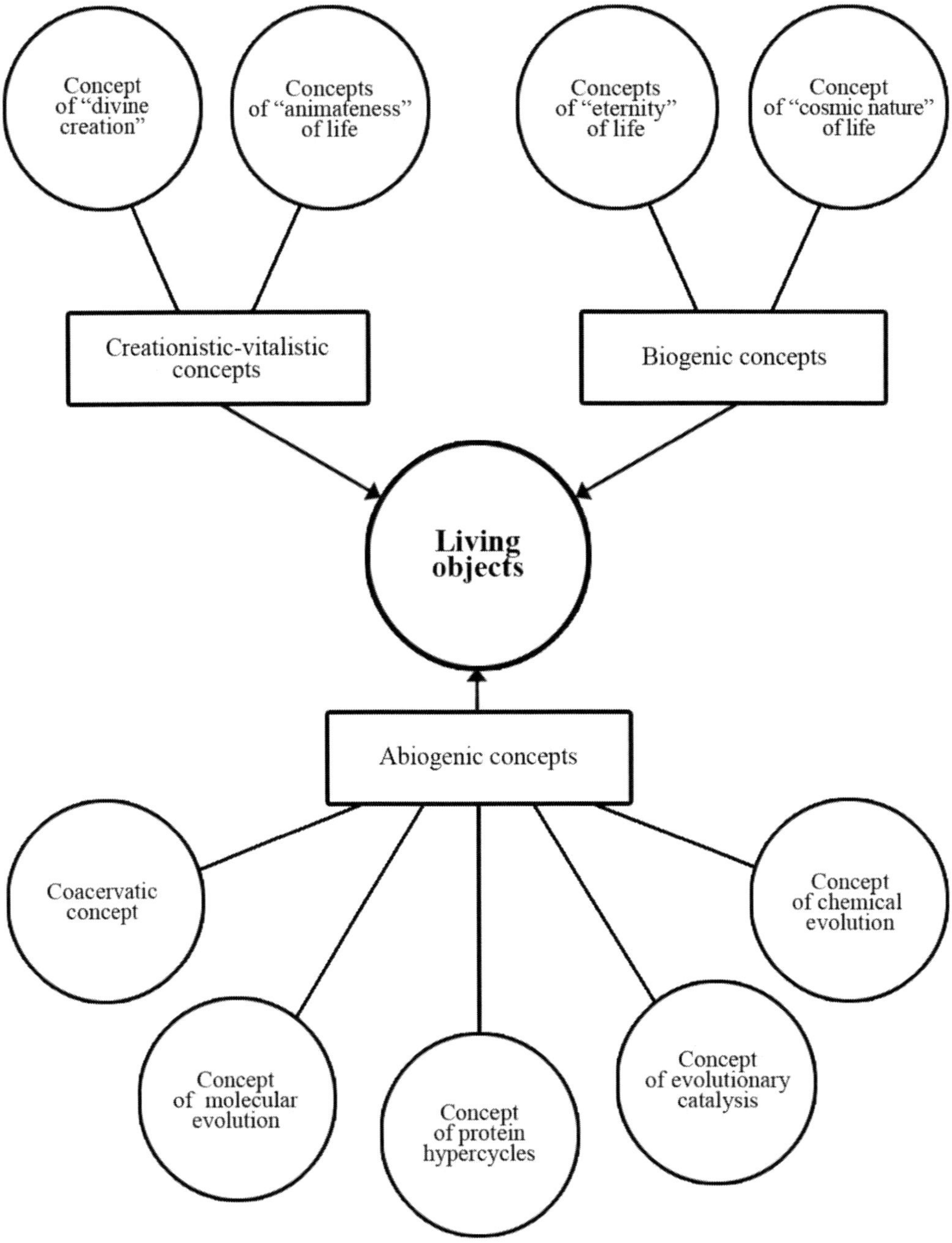

Cognition principles of life

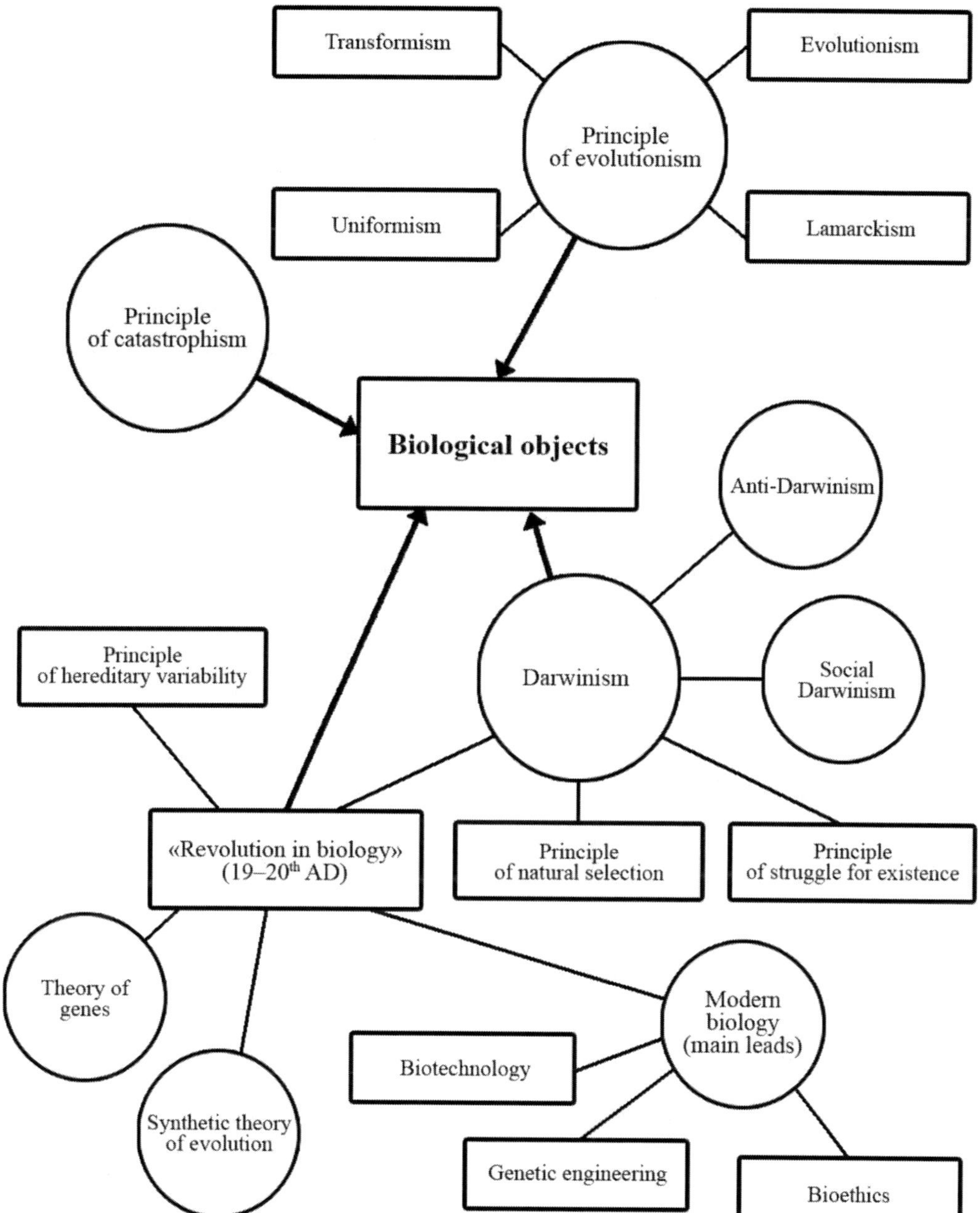

Modern structure of the general and applied ecology

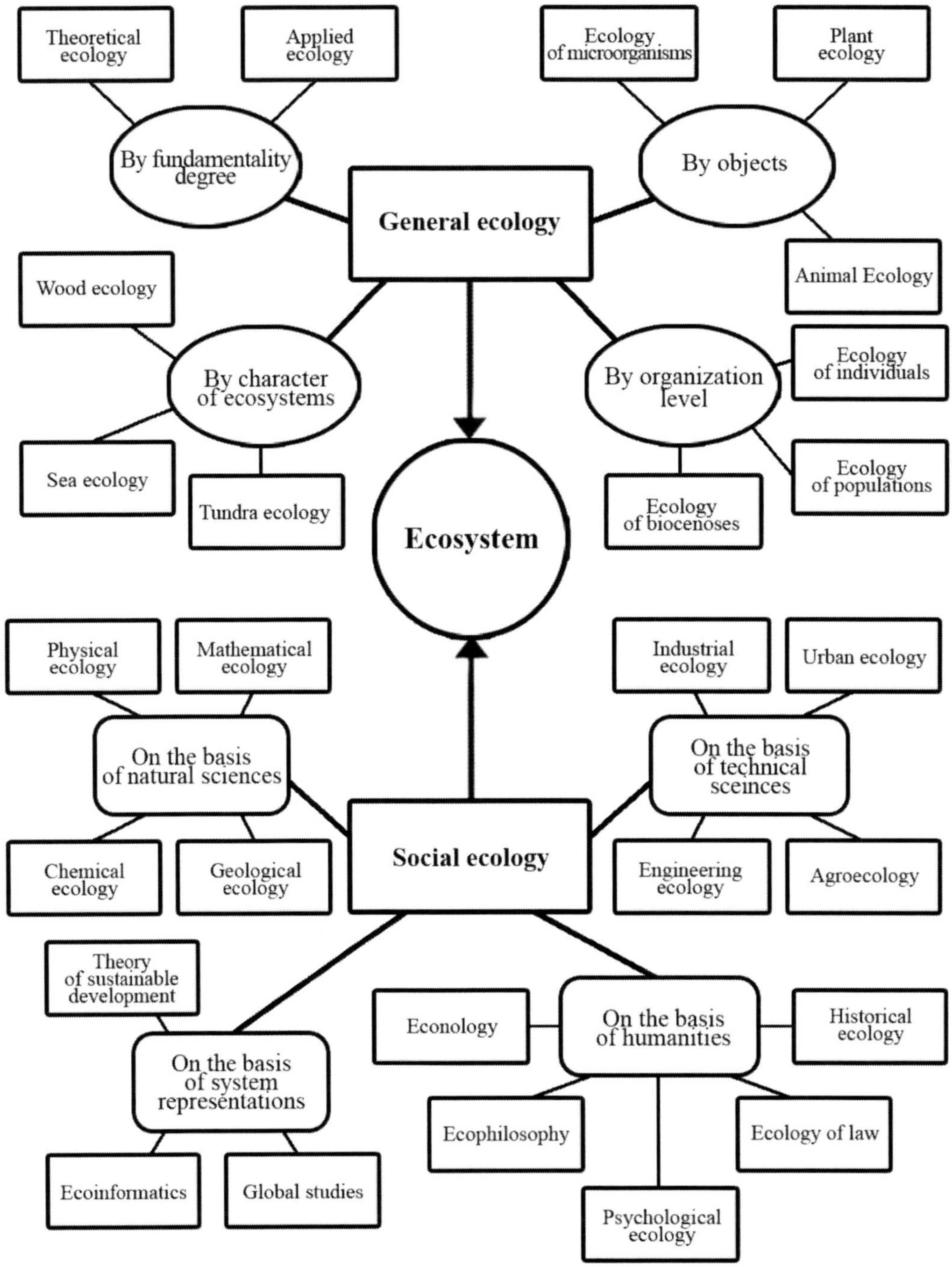

APPENDIX 2. A CALENDAR OF GLOBAL EVOLUTIONARY EVENTS

Calendar of the Precambrian Era on a 1 : 2·10[10] timescale

New Year	Big (hot) Bang
June	Emergence of the Milky Way—our Galaxy.
September	Emergence of the Solar system and formation of the planet Earth.
October	First live organisms; most ancient of the known residual rocks; most ancient of the known fossil forms of microorganisms.
November	Spreading of oxygen-producing microflora and microfauna; sexual reproduction development; emergence of photosynthesizing plants; first cells with nuclei (eukaryotes) capable of mitosis.
New Year	Formation of the atmosphere with high oxygen content; intense eruptions of volcanoes.
December	Formation of miotico-sexual reproduction.
Mid-December	Development of heterotrophic unicellular organisms; emergence of first multicellular organisms, the beginning of macroscopic life

Calendar of the Cambrian and Post-Cambrian Era on a 1 : 2·10[10] timescale

20.12	Emergence of invertebrates.
21.12	First oceanic plankton; blood of trilobites.
22.12	Ordovician (period): first vertebrates (fishes).
23.12	Silurian (period): Rhyniopsida occupy the land.
24.12	Devon (period): first insects; animals occupy the land; first amphibians; flying insects.
25.12	Carboniferous (period): first ferns; first reptiles.
26.12	Permian (period): first dinosaurs.
27.12	Triassic (period): first mammals.
28.12	Jurassic (period): first birds.
29.12	Cretaceous (period): first floral plants; extinction of dinosaurs.
30.12	Tertiary period: various mammals; first primacies; first hominids.
31.12	Emergence of Proconsul and Ramapithecus – ~ 14:00:00. First people – ~ 22:30:00. Usage of stone tools – ~ 23:00:00. Agriculture invention – ~ 13:59:00. First cities – ~ 23:59:30. Appearance of a written language – ~ 23:59:54. Bronze age – ~ 23:59:56. Iron Age – ~ 23:59:57. Euclidean geometry, Archimedean physics – ~ 23:59:59. Time reference mark – 24:00:00.
(New Year)	Introduction of zero and the decimal notation – ~ 00:00:01.
01.01	The Renaissance and modern science – ~ 00:00:02. The modern era – ~ 00:00:03.

Calendar of the terrestrial evolution on a 1 : 10^9 timescale

15–20 years ago	Big Bang (the exact date is unknown).
10 years ago	Emergence of our Galaxy.
6 years ago	Formation of the Solar system.
5 years ago	Earth emergence.
4 years ago	Emergence of the first forms of life.
$3^1/_2$ years ago	Formation of the most ancient of known fossils: bacteria, blue algae (photosynthesis!)
2 years 9 months ago	Beginning of the Archean (Archean Eon); beginning of the era of algae; emergence of free O_3 in the atmosphere.
2 years 3 months ago	Sexual reproduction development.
2 years ago	Beginning of the Proterozoic; formation of the big continent Gondwana.
1 year 10 months ago	Enhanced spreading of unicellular organisms.
1 year 9 months ago	Divergence between the evolutionary paths of prokaryotes and eukaryotes.
1 year 8 months ago	First multicellular organisms.
6 months ago	Cambrian, the dominating fauna: sea invertebrates; the area where the modern Morocco is located was at the South Pole.
5 months ago	Ordovician: first vertebrates.
$4^3/_4$ months ago	Silurian: merge of the microcontinents into a big continent Laurasia; formation of the Caledonian folding; Greenland and Tierra del Fuego are on the equator; first coral reeves and first fishes; plants occupy the land.
$4^1/_2$ months ago	Devonian: global temperature fall; Rio de Janeiro located at the Namibian coast is at the South Pole. In flora, Rhyniopsida concede domination to algae; in fauna, first overland vertebrates (amphibians) develop from fishes (Crossopterygii). Appearance of first insects. Animals occupy the land.
4 months ago	Carboniferous: further global temperature fall; Lavrasia and Gondwana merge into the pro-continent Pangaea; Variscan folding; Botswana at the South Pole; Siberia is close to the North Pole; coal woods; huge dragonflies; first reptiles.
$3^1/_2$ months ago	Permian: global temperature minimum; Antarctica at the South Pole; trilobites die out; Gymnosperms become dominating flora; first dinosaurs; first predecessors of mammals.
$2^1/_2$ months ago	Triassic: global temperature increase; disintegration of Pangaea into 4 continents: – Eurasia/North America, – Southern America/Africa, – India, –Antarctica/Australia. Kamchatka lies at the North Pole; reptiles break the domination of amphibians.
2 months ago	Jurassic: further global temperature increase; blossoming of huge dinosaurs; the first bird Archaeopteryx.

7 weeks ago	Cretaceous: global temperature maximum; the beginning of separation of South America from Africa and North America from Europe; formation of the Cordilleras; Africa, India, and Antarctica/Australia drift north, Africa meets Eurasia; formation of the Alpine folding; dinosaurs die out; the angiosperms start to dominate over Gymnosperm.
Last three weeks	Tertiary period: temperature globally decreases; South America separates from Africa; after the northwest part, the northeast part of Africa adjoins the Euroasian continent; the Mittelgebirgsschwelle is formed; Australia separates from Antarctica; domination of mammals begins; first primates.
Yesterday	Quaternary Period: further increase of the average global temperature; approach of "fast" temperature oscillations (ice ages); development of the man.

Calendar of the human evolution on a 1 : 10^9 timescale

7–11 days ago	Evolution way divergence of the Ponginae (the predecessors of orangutans, gorillas, and chimpanzee) and the Hominidae family (anthropoids).
5 days ago	Emergence of Ramapithecus, the most ancient of the known hominids; biped walking; use of suitable natural subjects as "tools of labor"; hunting for big animals.
2 days ago	Beginning of a transitional animal—man phase; appearance of Australopithecus (the man's ancestors) as direct predecessors of the hominids (the man); distinctly expressed walking on two feet; use of subjects either unwrought (natural) or processed by means of corporal bodies as an instrument of labor; hunting for large animals; collecting; life in herds.
1 day ago	End of the transitional animal—man phase; processing of natural subjects (e.g. sticks, stones) by means of other subjects as individual and then group-specific work (labor activity); beginning of public consciousness; a primitive human herd.
20 hours ago	Emergence of habilines (first people or Pithecanthropus); most ancient of the known stone tools (Oldowan stone tools); first human hominids: Homo habilis.
17 hours ago	Emergence of Homo erectus (primitive people); a manual chisel and an elementary tool from upholstered stones; fire usage; first huts; hunting-collecting economy; cooperation and work distribution in primitive groups; language emergence.
9 hours ago	Beginning of ice ages (cooling and drowning follow every 30 minutes); extinction of Australopithecuses and habilines; spreading of hominids; standardization of labor instruments and weapon.
2 hours ago	Emergence of Neanderthals (ancient people); processed instruments of labor from sharp-edged stones (tips, clips, stone knifes, etc.); tips from stone nuclei and chips; combined tools of stone and wood; bone tools; fire getting; life in huts, a human herd.
80 minutes ago	Beginning of the last ice age; fire usage; domination of more complex equipment; emergence of many tools of various purposes; century of the Neanderthal man.

Calendar of the human evolution on a 1 : 10^9 timescale (Continued)

30 minutes ago	Settlements of Australians; tips, harpoons, fire getting, burial, the cult of shadows, cannibalism.
8 minutes ago	End of the last ice age.
6 minutes ago	Mass production of semi-finished products; numerous special tools; art; magic of hunting and fertility; appearance of the modern man; differentiation of races; beginning of a profound phase of division of labor; early patrimonial system; first towns.

Calendar of the human society evolution on a 1 : 10^9 time scale

5 minutes ago	Neolithic civilization; first dynasties in Sumer, Ebla, and Egypt.
$2^1/_2$ minutes ago	Astronomy emergence; first class (slaveholding) society.
$2^1/_4$ minutes ago	Akkadian Empire; alphabet invention.
2 minutes ago	Middle Kingdom of Egypt.
$1^3/_4$ minutes ago	Trojan war; bronze age.
$1^1/_2$ minutes ago	Foundation of Carthago; Iron Age.
A minute ago	Beginning of the new era (AD).
45 seconds ago	Death of the Roman Empire; introduction of decimal notation with zero; feudalism blossoming.
35 seconds ago	Empire of Charles the Great, King of the francs.
28 seconds ago	Gothic architecture emergence.
15 seconds ago	Columbus' discovery of America.
12 seconds ago	Galileo Galilee's foundation of modern physics.
10 seconds ago	First capitalist state.
2 seconds ago	First socialist state; explosive blossoming of science and engineering.
A second ago	Second World War end

APPENDIX 3. NOBEL LAUREATES

List of Nobel laureates in Physics

1901	*Wilhelm Conrad Röntgen (Germany) (27.03.1845–10.02.1923)*	"in recognition of the extraordinary services he has rendered by the discovery of the remarkable rays subsequently named after him"
1902	*Hendrik Lorentz (Netherlands) (18.07.1853–4.02.1928), Pieter Zeeman (Netherlands) (25.05.1865–9.10.1943)*	"in recognition of the extraordinary service they rendered by their researches into the influence of magnetism upon radiation phenomena"
1903	*$^1/_2$–Antoine Henri Becquerel (France)(15.12.1852–25.08.1908)* *$^1/_2$–Pierre Curie (France) (1859–1906)and Marie Curie (France) (7.11.1867–4.07.1934)*	"[for] his discovery of spontaneous radioactivity" "[for] their joint researches on the radiation phenomena discovered by Professor Henri Becquerel"
1904	*John William Lord Rayleigh (UK) (12.11.1842–30.06.1919)*	"for his investigations of the densities of the most important gases and for his discovery of argon in connection with these studies"

1905	*Philipp Eduard Anton von Lenard (Austria–Hungary) (7.06.1862–20.05.1947)*	"for his work on cathode rays"
1906	*Joseph John Thomson (UK) (18.12.1856–30.08.1940)*	"[for] his theoretical and experimental investigations on the conduction of electricity by gases"
1907	*Albert Abraham Michelson (USA) (19.12.1852–09.051931)*	"for his optical precision instruments and the spectroscopic and metrological investigations carried out with their aid"
1908	*Gabriel Lippmann (France) (16.08.1845–13.07.1921)*	"for his method of reproducing colours photographically based on the phenomenon of interference"
1909	*Guglielmo Marconi (Italy) (25.04.1874–20.07.1937) and Karl Ferdinand Braun (Germany) (06.06.1850–20.04.1918)*	"[for] their contributions to the development of wireless telegraphy"
1910	*Johannes Diderik van der Waals (Netherlands) (23.11.1837–08.03.1923)*	"for his work on the equation of state for gases and liquids"
1911	*Wilhelm Wien (Germany) (13.01.1864–30.08.1928)*	"for his discoveries regarding the laws governing the radiation of heat"
1912	*Nils Gustaf Dalén (Sweden) (30.11.1869–09.12.1937)*	"for his invention of automatic valves designed to be used in combination with gas accumulators in lighthouses and buoys"
1913	*Heike Kamerlingh-Onnes (Netherlands) (21.09.1853–21.02.1926)*	"for his investigations on the properties of matter at low temperatures which led, inter alia, to the production of liquid helium"
1914	*Max von Laue (Germany) (09.10.1879–24.04.1960)*	"For his discovery of the diffraction of X-rays by crystals"
1915	*William Henry Bragg (UK) (02.07.1862–12.03.1942)andu William Lawrence Bragg (UK) (31.03.1890–01.07.1971)*	"For their services in the analysis of crystal structure by means of X-rays"
1916	*Not awarded*	
1917	*Charles Glover Barkla (UK) (07.06.1877–23.10.1944) (awarded in 1918)*	"For his discovery of the characteristic Röntgen radiation of the elements"
1918	*Max Karl Ernest Ludwig Planck (Germany)(23.04.1858–04.10.1947) (awarded in 1919)*	"[for] the services he rendered to the advancement of Physics by his discovery of energy quanta"
1919	*Johannes Stark (Germany) (15.04.1874–21.06.1957)*	"for his discovery of the Doppler effect in canal rays and the splitting of spectral lines in electric fields"
1920	*Charles Édouard Guillaume (France) (15.02.1861–13.06.1938)*	"[for] the service he has rendered to precision measurements in physics by his discovery of anomalies in nickel-steel alloys"
1921	*Albert Einstein (Germany) (14.03.1879–18.04.1955) (awarded in 1922)*	"for his services to Theoretical Physics, and especially for his discovery of the law of the photoelectric effect"
1922	*Niels Bohr (Denmark) (14.03.1885–18.11.1962)*	"for his services in the investigation of the structure of atoms and of the radiation emanating from them"
1923	*Robert Andrews Millikan (USA) (22.03.1868–19.12.1953)*	"for his work on the elementary charge of electricity and on the photoelectric effect"
1924	*Manne Siegbahn (Sweden) (03.12.1886–26.09.1978) (awarded in 1925)*	"for his discoveries and research in the field of X-ray spectroscopy"

List of Nobel laureates in Physics (Continued)

1925	*James Franck (Germany) (26.08.1882–21.05.1964) and Gustav Hertz (Germany) (1887–1975) (awarded in 1926)*	"for their discovery of the laws governing the impact of an electron upon an atom"
1926	*Jean Baptiste Perrin (France) (30.09.1870–17.04.1942)*	"for his work on the discontinuous structure of matter, and especially for his discovery of sedimentation equilibrium"
1927	*Arthur Holly Compton (USA) (10.09.1892–15.03.1962)* *Charles Thomson Rees Wilson (Scotland) (14.02.1869–15.11.1959)*	"for his discovery of the effect named after him" "for his method of making the paths of electrically charged particles visible by condensation of vapour"
1928	*Owen Willans Richardson (UK) (26.04.1892–15.02.1959)*	"for his work on the thermionic phenomenon and especially for the discovery of the law named after him"
1929	*Prince Louis-Victor Pierre Raymond de Broglie (France) (1892–1987)*	"for his discovery of the wave nature of electrons"
1930	*Chandrasekhara Venkata Raman (India) (7.11.1888–21.11.1970)*	"for his work on the scattering of light and for the discovery of the effect named after him"
1931	*Not awarded*	
1932	*Werner Heisenberg (Germany) (5.12.190–1.02.1976) (awarded in 1933)*	"for the creation of quantum mechanics, the application of which has, inter alia, led to the discovery of the allotropic forms of hydrogen"
1933	*Erwin Schrödinger (Austria) (12.08.1887–04.01.1961) and Paul Adrien Morris Dirac (UK) (08.08.1902–20.10.1984)*	"for the discovery of new productive forms of atomic theory"
1934	*Not awarded*	
1935	*James Chadwick (UK) (20.10.1891–24.07.1974)*	"for the discovery of the neutron"
1936	*Victor Francis Hess (USA) (24.06.1883–17.12.1964)* *Carl David Anderson (USA) (03.09.1905–11.01.1991)*	"for his discovery of cosmic radiation" «for his discovery of the positron»
1937	*Clinton Joseph Davisson (USA) (22.10.1881–01.02.1958) and George Paget Thomson (UK) (03.05.1892–10.09.1975)*	"for their experimental discovery of the diffraction of electrons by crystals"
1938	*Enrico Fermi (USA) (29.09.1901–28.11.1954)*	"for his demonstrations of the existence of new radioactive elements produced by neutron irradiation, and for his related discovery of nuclear reactions brought about by slow neutrons"
1939	*Ernest Orlando Lawrenc (USA) (08.08.1901–27.08.1958)*	"for the invention and development of the cyclotron and for results obtained with it, especially with regard to artificial radioactive elements"
1940	*Not awarded*	
1941	*Not awarded*	
1942	*Not awarded*	
1943	*Otto Stern (Germany) (17.02.1888–17.08.1969) (awarded in 1944)*	"for his contribution to the development of the molecular ray method and his discovery of the magnetic moment of the proton"

1944	*Isidor Isaac Rabi (USA) (29.07.1898–11.01.1988)*	"for his resonance method for recording the magnetic properties of atomic nuclei"
1945	*Wolfgang Pauli (Switzrland) (25.04.1900–15.12.1958)*	"for the discovery of the Exclusion Principle, also called the Pauli principle"
1946	*Percy Williams Bridgman (USA) (21.04.1882–20.08.1961)*	"for the invention of an apparatus to produce extremely high pressures, and for the discoveries he made there within the field of high pressure physics"
1947	*Edward Victor Appleton (UK) (06.09.1892–21.04.1965)*	"for his investigations of the physics of the upper atmosphere especially for the discovery of the so-called Appleton layer"
1948	*Patrick Maynard Stuart Blackett (UK) (18.11.1897–13.07.1974)*	"for his development of the Wilson cloud chamber method, and his discoveries therewith in the fields of nuclear physics and cosmic radiation"
1949	*Hideki Yukawa (Japan) (23.01.1907–08.09.1981)*	"for his prediction of the existence of mesons on the basis of theoretical work on nuclear forces"
1950	*Cecil Frank Powell (UK) (05.12.1903–09.08.1969)*	"for his development of the photographic method of studying nuclear processes and his discoveries regarding mesons made with this method"
1951	*John Douglas Cockcroft (UK) (27.05.1897–18.09.1967) and Ernest Thomas Sinton Walton (Ireland) (06.10.1903–25.06.1995)*	"for their pioneer work on the transmutation of atomic nuclei by artificially accelerated atomic particles"
1952	*Felix Bloch (USA) (23.10.1905–10.09.1983) and Edward Mills Purcell (USA) (30.08.1912–07.03.1997)*	"for their development of new methods for nuclear magnetic precision measurements and discoveries in connection therewith"
1953	*Frits Zernike (Netherlands) (16.07.1888–10.03.1966)*	"for his demonstration of the phase contrast method, especially for his invention of the phase contrast microscope"
1954	*Max Born (Germany) (11.12.1882–05.01.1970)*	"for his fundamental research in quantum mechanics, especially for his statistical interpretation of the wavefunction"
	Walther Bothe (West Germany) (08.01.1891–08.02.1957)	"for the coincidence method and his discoveries made therewith"
1955	*Willis Eugene Lamb (USA) (12.07.1913–21.05.2008)*	"for his discoveries concerning the fine structure of the hydrogen spectrum"
	Polykarp Kusch (USA) (26.01.1911–20.03.1993)	"for his precision determination of the magnetic moment of the electron"
1956	*William Bradford Shockley (USA) (13.02.1908–12.08.1989), John Bardeen (USA) (23.05.1908–30.01.1991) and Walter Houser Brattain (USA) (10.02.1902–13.10.1987)*	"for their researches on semiconductors and their discovery of the transistor effect"
1957	*Chen Ning Yang (China) (b. 1.10.1922) and Tsung-Dao Lee (China) (b. 24.11.1926)*	"for their penetrating investigation of the so-called parity laws which has led to important discoveries regarding the elementary particles"
1958	*Pavel Alekseyevich Cherenkov (USSR) (28.07.1904–06.01.1990), Ilya Mikhailovich Frank (USSR) (10(23).10.1908–22.06.1990)and Igor Yevgenyevich Tamm (USSR) (26.06.(08.07.)1895–12.04.1971)*	"for the discovery and the interpretation of the Cherenkov effect"

List of Nobel laureates in Physics (Continued)

1959	*Emilio Gino Segrè (USA) (01.02.1905–22.04.1989) and Owen Chamberlain (USA) (10.07.1920–28.02.2006)*	"for their discovery of the antiproton"
1960	*Donald Arthur Glaser (USA) (b. 21.09.1926)*	"for the invention of the bubble chamber"
1961	*Robert Hofstadter (USA) (05.02.1915–17.11.1990)*	"for his pioneering studies of electron scattering in atomic nuclei and for his thereby achieved discoveries concerning the structure of the nucleons"
	Rudolf Ludwig Mössbauer (West Germany) (b. 31.01.1929)	"for his researches concerning the resonance absorption of gamma radiation and his discovery in this connection of the effect which bears his name
1962	*Lev Davidovich Landau (USSR) (9(22).01.1908–01.04.1968)*	"for his pioneering theories for condensed matter, especially liquid helium"
1963	*½ – Eugene Paul Wigner (USA) (17.11.1902–01.01.1995)*	"for his contributions to the theory of the atomic nucleus and the elementary particles, particularly through the discovery and application of fundamental symmetry principles"
	½ – Maria Goeppert-Mayer (USA) (28.06.1906–20.02.1972) and J. Hans D. Jensen (West Germany) (25.06.1907–11.02.1973)	"for their discoveries concerning nuclear shell structure"
1964	*½ – Nicolay Gennadiyevich Basov (USSR) (14.12.1922–01.07.2001), Alexander Mikhailovich Prokhorov (USSR) (11.07.1916–08.01.2002)* *½ – Charles Hard Townes (USA) (b. 28.07.1915)*	"for fundamental work in the field of quantum electronics, which has led to the construction of oscillators and amplifiers based on the maser-laser principle"
1965	*Sin-Itiro Tomonaga (Japan) (31.03.1906–08.07.1979), Julian Schwinger (USA) (12.02.1918–16.07.1994) and Richard Phillips Feynman (USA) (11.05.1918–15.02.1988)*	"for their fundamental work in quantum electrodynamics, with deep-ploughing consequences for the physics of elementary particles"
1966	*Alfred Kastler (France) (03.05.1902–07.01.1984)*	"for the discovery and development of optical methods for studying Hertzian resonances in atoms"
1967	*Hans Albrecht Bethe (USA) (02.07.1906–06.03.2005)*	"for his contributions to the theory of nuclear reactions, especially his discoveries concerning the energy production in stars"
1968	*Luis Walter Alvarez (USA) (13.06.1911–01.09.1988)*	"for his decisive contributions to elementary particle physics, in particular the discovery of a large number of resonance states, made possible through his development of the technique of using hydrogen bubble chamber and data analysis"
1969	*Murray Gell-Mann (USA) (b. 15.09.1929)*	"for his contributions and discoveries concerning the classification of elementary particles and their interactions"

1970	*Hannes Olof Gösta Alfvén (Sweden) (30.05.1908–02.04.1995)* *Louis Néel (France) (22.11.1904–17.11.2000)*	"for fundamental work and discoveries in magneto-hydrodynamics with fruitful applications in different parts of plasma physics» "for fundamental work and discoveries concerning antiferromagnetism and ferrimagnetism which have led to important applications in solid state physics"
1971	*Dennis Gabor (UK) (05.06.1900–09.02.1979)*	"for his invention and development of the holographic method"
1972	*John Bardeen (USA) (23.05.1908–30.01.1991), Leon Neil Cooper (USA) (b. 28.02.1930) and John Robert Schrieffer (USA) (b. 31.05.1931)*	"for their jointly developed theory of superconductivity, usually called the BCS-theory"
1973	*½ – Brian David Josephson (UK) (b. 04.01.1940)* *½ – Leo Esaki (Japan) (b. 12.03.1925) and Ivar Giaever (Norway) (b. 05.04.1929)*	"for his theoretical predictions of the properties of a supercurrent through a tunnel barrier, in particular those phenomena which are generally known as the Josephson effect" "for their experimental discoveries regarding tunneling phenomena in semiconductors and superconductors, respectively"
1974	*Martin Ryle (UK) (27.09.1918–14.10.1984) and Antony Hewish (UK) (b. 11.05.1924)*	"for their pioneering research in radio astrophysics"
1975	*Aage Niels Bohr (Denmark) (b. 19.06.1922), Ben Roy Mottelson (USA) (09.07.1926) and Leo James Rainwater(USA)*	"for the discovery of the connection between collective motion and particle motion in atomic nuclei and the development of the theory of the structure of the atomic nucleus based on this connection"
1976	*Burton Richter (USA) (b. 22.03.1931) and Samuel Chao Chung Ting (USA) (b. 27.01.1936)*	"for their pioneering work in the discovery of a heavy elementary particle of a new kind"
1977	*Philip Warren Anderson (USA) (b. 13.12.1923), Nevill Francis Mott (UK) (30.09.1905–08.08.1996) and John Hasbrouck Van Vleck (USA) (13.03.1899–27.10.1980)*	"for their fundamental theoretical investigations of the electronic structure of magnetic and disordered systems"
1978	*½ – Pyotr Leonidovich Kapitsa (USSR) (26.06.(09.07.)1894–08.04.1984)* *½ – Arno Allan Penzias (USA) (b. 26.04.1933) and Robert Woodrow Wilson (USA) (b. 10.01.1936)*	"for his basic inventions and discoveries in the area of low-temperature physics" "for their discovery of cosmic microwave background radiation"
1979	*Sheldon Lee Glashow (USA) (b. 05.12.1932), Abdus Salam (Pakistan) (29.01.1926–21.11.1996) and Steven Weinberg (USA) (b. 1933)*	"for their contributions to the theory of the unified weak and electromagnetic interaction between elementary particles, including, inter alia, the prediction of the weak neutral current"
1980	*James Watson Cronin (USA) (b. 29.09.1931) and Val Logsdon Fitch (USA) (b. 10.03.1923)*	"or the discovery of violations of fundamental symmetry principles in the decay of neutral K-mesons"
1981	*½ – Nicolaas Bloembergen (Netherlands) (b. 20.03.1920) and Arthur Leonard Schawlow (USA) (05.05.1921–28.04.1999)*	"for their contribution to the development of laser spectroscopy"

List of Nobel laureates in Physics (Continued)

	½ – Kai Manne Börje Siegbahn (Sweden) (20.04.1918-20.07.2007)	"for his contribution to the development of high-resolution electron spectroscopy"
1982	*Kenneth G. Wilson (USA) (b. 08.06.1936)*	"for his theory for critical phenomena in connection with phase transitions"
1983	*Subrahmanyan Chandrasekhar (USA) (19.10.1910–21.08.1995)*	"for his theoretical studies of the physical processes of importance to the structure and evolution of the stars"
	William Alfred Fowler (USA) (09.08.1911–14.03.1995)	"for his theoretical and experimental studies of the nuclear reactions of importance in the formation of the chemical elements in the universe"
1984	*Carlo Rubbia (Italy) (b. 31.03.1934) and Simon van der Meer (Netherlands) (b. 24.11.1925)*	"for their decisive contributions to the large project, which led to the discovery of the field particles W and Z, communicators of weak interaction"
1985	*Klaus von Klitzing (West Germany) (b. 28.06.1943)*	"for the discovery of the quantized Hall effect"
1986	*½ – Ernst Ruska (West Germany) (25.12.1906–27.05.1988)*	"for his fundamental work in electron optics, and for the design of the first electron microscope"
	½ – Gerd Binnig (West Germany) (b. 1947) and Heinrich Rohrer (Switzerland) (b. 06.06.1933)	"for their design of the scanning tunneling microscope"
1987	*Johannes Georg Bednorz (West Germany) (b. 16.05.1950) and Karl Alexander Müller (Switzerland) (b. 20.04.1927)*	"for their important break-through in the discovery of superconductivity in ceramic materials"
1988	*Leon Max Lederman (USA) (b. 15.07.1922), Melvin Schwartz (USA) (02.11.1932–28.08.2006) and Jack Steinberger (USA) (b. 25.05.1921)*	"for the neutrino beam method and the demonstration of the doublet structure of the leptons through the discovery of the muon neutrino"
1989	*½ – Norman Foster Ramsey (USA) (b. 27.08.1915)*	"for the invention of the separated oscillatory fields method and its use in the hydrogen maser and other atomic clocks"
	½ – Hans Georg Dehmelt)(Germany) (b. 09.09.1922)andu Wolfgang Paul (Austria)) (10.08.1913–07.12.1993)	"for the development of the ion trap technique"
1990	*Jerome I. Friedman (USA) (b. 28.03.1930), Henry Way Kendall (USA) (09.12.1926–15.02.1999) and Richard E. Taylor (UK) (02.11.1929)*	"for their pioneering investigations concerning deep inelastic scattering of electrons on protons and bound neutrons, which have been of essential importance for the development of the quark model in particle physics"
1991	*Pierre-Gilles de Gennes (France) (24.10.1932–18.05.2007)*	"for discovering that methods developed for studying order phenomena in simple systems can be generalized to more complex forms of matter, in particular to liquid crystals and polymers"
1992	*Georges Charpak (France) (b. 01.08.1924)*	"for his invention and development of particle detectors, in particular the multiwire proportional chamber"
1993	*Russell Alan Hulse (USA) (b. 28.11.1950) and Joseph Hooton Taylor,Jr. (USA) (b. 29.03.1941)*	"for the discovery of a new type of pulsar, a discovery that has opened up new possibilities for the study of gravitation"

1994	*Bertram Brockhouse (Canada) (15.07.1918–13.10.2003)* *Clifford Glenwood Shull (USA) (23.09.1915–31.03.2001)*	"for the development of neutron spectroscopy" "for the development of the neutron diffraction technique"
1995	*Martin Lewis Perl (USA) (b. в 1927)* *Frederick Reines (USA) (16.03.1918–26.08.1998)*	"for the discovery of the tau lepton" "for the detection of the neutrino"
1996	*David Morris Lee (USA) (b. 20.01.1931), Douglas D. Osheroff (USA) (b. 01.08.1945) and Robert Coleman Richardson (USA) (b. 26.06.1937)*	"for their discovery of superfluidity in helium-3"
1997	*Steven Chu (USA) (b. 28.02.1948), Claude Cohen-Tannoudji (France) (b. 01.04.1933) and William Daniel Phillips (USA) (b. 05.11.1948)*	"for development of methods to cool and trap atoms with laser light"
1998	*Robert B. Laughlin (USA) (b. 01.11.1950), Horst Ludwig Störmer (Germany) (b. 06.04.1949) and Daniel Chee Tsui (USA) (b. 28.02.1939)*	"for their discovery of a new form of quantum fluid with fractionally charged excitations"
1999	*Gerard 't Hooft (Netherlands) (b. 05.07.1946) and Martinus J. G. Veltman (USA) (b. 27.06.1931)*	"for elucidating the quantum structure of electroweak interactions in physics"
2000	*½ – Zhores Ivanovich Alferov (Russia) (b. 15.03.1930) and Herbert Kroemer (USA) (b. 25.08.1928)* *½ – Jack St. Clair Kilby (USA) (08.11.1923-20.06.2005)*	"for developing semiconductor heterostructures used in high-speed- and optoelectronics" "for his part in the invention of the integrated circuit"
2001	*Eric Allin Cornell (USA) (b. 19.12.1961), Wolfgang Ketterle (Germany) (21.10.1957) and Carl Edwin Wieman (USA) (b. 26.03.1951)*	"for the achievement of Bose-Einstein condensation in dilute gases of alkali atoms, and for early fundamental studies of the properties of the condensates"
2002	*½ – Raymond Davis, Jr. (USA) (14.10.1914–31.05.2006) and Masatoshi Koshiba (Japan) (b. 19.09.1926)* *½ – Riccardo Giacconi (Italy) (b. 06.10.1931)*	"for pioneering contributions to astrophysics, in particular for the detection of cosmic neutrinos" "for pioneering contributions to astrophysics, which have led to the discovery of cosmic X-ray sources"
2003	*Alexei Alexeyevich Abrikosov (Россия) (b. 25.06.1928), Vitaly Lazarevich Ginzburg (Россия) (b. 04.10.1916), Anthony James Leggett (UK) (b. 26.03.1938)*	"for pioneering contributions to the theory of superconductors and superfluids"
2004	*David J. Gross (USA) (b. 19.02.1941), Hugh David Politzer (USA) (b. 31.08.1949) and Frank Wilczek (USA) (15.05.1951)*	"for the discovery of asymptotic freedom in the theory of the strong interaction"
2005	*½ – Roy J. Glauber (USA) (b. 01.09.1925)* *½ – John L. Hall (USA) (b. 21.08.1934) and Theodor W. Hänsch (Germany) (b. 30.10.1941)*	"for his contribution to the quantum theory of optical coherence" "for their contributions to the development of laser-based precision spectroscopy, including the optical frequency comb technique"

List of Nobel laureates in Physics (Continued)

2006	*John C. Mather (USA) (b. 07.03.1938) and George F. Smoot (USA) (b. 20.02.1945)*	"for their discovery of the blackbody form and anisotropy of the cosmic microwave background radiation"
2007	*Albert Fert (France) (b. 07.03.1938) and Peter Grünberg (Germany) (b. 18.05.1939)*	"for the discovery of giant magnetoresistance"
2008	*½ – Yoichiro Nambu (USA) (b. 18.01.1921)*	"for the discovery of the mechanism of spontaneous broken symmetry in subatomic physics"
	½ – Makoto Kobayashi (Japan) (b. 07.04.1944) and Toshihide Maskawa (Japan) (b. 07.02.01940)	"for the discovery of the origin of the broken symmetry which predicts the existence of at least three families of quarks in nature"
2009	*½–Charles K. Kao (China) (b. 04.11.1933)*	"for groundbreaking achievements concerning the transmission of light in fibers for optical communication"
	¼ – Willard S. Boyle (USA) (b. 08.07.1924) and George E. Smith (USA) (26.03.1840-19.08.18760)	"for the invention of an imaging semiconductor circuit–the CCD sensor"
2010	*Andre Geim (Netherlands) (b. 21.10.1958) and Konstantin Novoselov (UK) (b. 23.08.1974)*	"for groundbreaking experiments regarding the two-dimensional material graphene"
2011	*Saul Perlmutter (USA) (b. 22.09.1959), Brian P. Schmidt (USA) (b.24.02.1967) and Adam G. Riess (USA) (b. 16.12.1969)*	"for the discovery of the accelerating expansion of the Universe through observations of distant supernovae"

List of Nobel laureates in Chemistry

1901	*Jacobus Henricus van 't Hoff (Netherlands) (30.08.1852–01.03.1911)*	"[for his] discovery of the laws of chemical dynamics and osmotic pressure in solutions"
1902	*Hermann Emil Fischer (Germany) (09.10.1852–15.07.1919)*	"[for] his work on sugar and purine syntheses"
1903	*Svante August Arrhenius (Sweden) (19.02.1859–02.10.1927)*	"[for] his electrolytic theory of dissociation"
1904	*Sir William Ramsay (UK) (02.10.1852–23.07.1916)*	"[for his] discovery of the inert gaseous elements in air, and his determination of their place in the periodic system"
1905	*Johann Friedrich Wilhelm Adolf von Baeyer (Germany) (31.10.1835–20.08.1917)*	"[for] the advancement of organic chemistry and the chemical industry, through his work on organic dyes and hydroaromatic compounds"
1906	*Henri Moissan (France) (28.09.1852–20.02.1907)*	"[for his] investigation and isolation of the element fluorine, and for [the] electric furnace called after him"
1907	*Eduard Buchner (Germany) (20.05.1860–13.08.1917)*	"for his biochemical researches and his discovery of cell-free fermentation"
1908	*Ernest Rutherford (UK) (30.08.1871–19.10.1937)*	"for his investigations into the disintegration of the elements, and the chemistry of radioactive substances"

1909	*Wilhelm Ostwald (Germany) (02.09.1853–04.04.1932)*	"[for] his work on catalysis and for his investigations into the fundamental principles governing chemical equilibria and rates of reaction"
1910	*Otto Wallach (Germany) (27.03.1847–26.02.1931)*	"[for] his services to organic chemistry and the chemical industry by his pioneer work in the field of alicyclic compounds"
1911	*Marie Curie, née Sklodowska (France) (07.11.1867–04.07.1934)*	"[for] the discovery of the elements radium and polonium, by the isolation of radium and the study of the nature and compounds of this remarkable element"
1912	*Victor Grignard (France) (06.05.1871-13.12.1935)* *Paul Sabatier (France) (05.11.1854–14.08.1941)*	"for the discovery of the [...] Grignard reagent" "for his method of hydrogenating organic compounds in the presence of finely disintegrated metals"
1913	*Alfred Werner (Switzerland) (12.12.1866–15.11.1919)*	"[for] his work on the linkage of atoms in molecules [...] especially in inorganic chemistry"
1914	*Theodore William Richards (Germany) (31.01.1868–02.04.1928)*	"[for] his accurate determinations of the atomic weight of a large number of chemical elements"
1915	*Richard Martin Willstätter (Germany) (13.08.1872–03.08.1942)*	"for his researches on plant pigments, especially chlorophyll"
1916	*Not awarded*	
1917	*Not awarded*	
1918	*Fritz Haber (Germany) (09.12.1868–29.01.1934)*	"for the synthesis of ammonia from its elements"
1919	*Not awarded*	
1920	*Walther Hermann Nernst (Germany) (25.06.1864–18.11.1941)*	"[for] his work in thermochemistry"
1921	*Frederick Soddy (UK) (02.09.1877–22.09.1956)*	"for his contributions to our knowledge of the chemistry of radioactive substances, and his investigations into the origin and nature of isotopes"
1922	*Francis William Aston (UK) (01.09.1877–20.11.1945)*	"for his discovery, by means of his mass spectrograph, of isotopes, in a large number of non-radioactive elements, and for his enunciation of the whole-number rule"
1923	*Fritz Pregl (UK) (03.09.1869–13.12.1930)*	"for his invention of the method of micro-analysis of organic substances"
1924	*Not awarded*	
1925	*Richard Adolf Zsigmond (Austria) (01.04.1865–23.09.1929)*	"for his demonstration of the heterogeneous nature of colloid solutions and for the methods he used"
1926	*Theodor Svedberg (Sweden) (30.08.1884–26.02.1971)*	"for his work on disperse systems"
1927	*Heinrich Otto Wieland (Germany) (04.06.1877–05.08.1957)*	"for his investigations of the constitution of the bile acids and related substances"
1928	*Adolf Otto Reinhold Windaus (Germany) (25.12.1876–09.06.1959)*	"[for] his research into the constitution of the sterols and their connection with the vitamins"

List of Nobel laureates in Chemistry (Continued)

1929	*Arthur Harden (UK) (12.10.1865–17.06.1940) and Hans Karl August Simon von Euler-Chelpin (*Sweden*) (15.02.1873–06.07.1964)*	"for their investigations on the fermentation of sugar and fermentative enzymes"
1930	*Hans Fischer (Germany) (27.07.1881–31.03.1945)*	"for his researches into the constitution of haemin and chlorophyll and especially for his synthesis of haemin"
1931	*Carl Bosch (Germany) (27.08.1874–26.04.1940) and Friedrich Bergius (Germany) (11.10.1884–31.03.1949)*	"[for] their contributions to the invention and development of chemical high pressure methods"
1932	*Irving Langmuir (USA) (31.01.1881–16.08.1957)*	"for his discoveries and investigations in surface chemistry"
1933	*Not awarded*	
1934	*Harold Clayton Urey (USA) (29.04.1893–05.01.1981)*	"for his discovery of heavy hydrogen"
1935	*Frédéric Joliot (France) (19.03.1900–14.08.1958) and Irène Joliot-Curie (France) (12.09.1897–17.03.1956)*	"[for] their synthesis of new radioactive elements"
1936	Petrus (Peter) Josephus Wilhelmus Debye (*Netherlands*) (24.03.1884–02.11.1966)	"[for his work on] molecular structure through his investigations on dipole moments and the diffraction of X-rays and electrons in gases"
1937	*Walter Norman Haworth (UK) (19.03.1883–19.03.1950)* *Paul Karrer (Switzerland)*	"for his investigations on carbohydrates and vitamin C" "for his investigations on carotenoids, flavins and vitamins A and B_2"
1938	*Richard Kuhn (Germany) (03.12.1900–31.07.1967)*	"for his work on carotenoids and vitamins"
1939	*Adolf Friedrich Johann Butenandt (Germany) (24.03.1903–18.01.1995)* *Leopold Ruzicka (Switzerland) (13.09.1887–26.09.1976)*	"for his work on sex hormones" "for his work on polymethylenes and higher terpenes"
1940	*Not awarded*	
1941	*Not awarded*	
1942	*Not awarded*	
1943	*George de Hevesy (Hungary) (01.08.1885–05.07.1966)*	"for his work on the use of isotopes as tracers in the study of chemical processes"
1944	*Otto Hahn (Germany) (08.03.1879–28.07.1968)*	"for his discovery of the fission of heavy nuclei"
1945	*Artturi Ilmari Virtanen (Finland) (15.01.1895–11.11.1973)*	"for his research and inventions in agricultural and nutrition chemistry, especially for his fodder preservation method"
1946	*½ – James Batcheller Sumner (USA) (19.11.1887–12.08.1955)* *½ – John Howard Northrop (USA) (05.07.1891–27.05.1987) and Wendell Meredith Stanley (USA) (16.08.1904–15.06.1971)*	"for his discovery that enzymes can be crystallized" "for their preparation of enzymes and virus proteins in a pure form"

1947	*Sir Robert Robinson (UK) (13.09.1886–24.02.1975)*	"for his investigations on plant products of biological importance, especially the alkaloids"
1948	*Arne Wilhelm Kaurin Tiselius (Sweden) (10.08.1902–29.10.1971)*	"for his research on electrophoresis and adsorption analysis, especially for his discoveries concerning the complex nature of the serum proteins"
1949	*William Francis Giauque (USA) (12.05.1895–28.03.1982)*	"for his contributions in the field of chemical thermodynamics, particularly concerning the behaviour of substances at extremely low temperatures"
1950	*Otto Paul Hermann Diels (West Germany) (23.01.1876–07.03.1954) and Kurt Alder (West Germany) (10.07.1902–20.06.1958)*	"for their discovery and development of the diene synthesis"
1951	*Edwin Mattison McMillan (USA) (18.09.1907–07.09.1991) and Glenn Theodore Seaborg (USA) (19.04.1912–25.02.1999)*	"for their discoveries in the chemistry of transuranium elements"
1952	*Archer John Porter Martin (UK) (01.03.1910–28.07.2002) and Richard Laurence Millington Synge (UK) (28.10.1914–18.08.1994)*	"for their invention of partition chromatography"
1953	*Hermann Staudinger (West Germany) (23.03.1881–08.09.1965)*	"for his discoveries in the field of macromolecular chemistry"
1954	*Linus Carl Pauling (USA) (28.02.1901–19.08.1994)*	"for his research into the nature of the chemical bond and its application to the elucidation of the structure of complex substances"
1955	*Vincent du Vigneaud (USA) (18.05.1901–11.12.1978)*	"for his work on biochemically important sulphur compounds, especially for the first synthesis of a polypeptide hormone"
1956	*Sir Cyril Norman Hinshelwood (UK) (19.06.1897–09.10.1967) and Nikolay Nikolaevich Semenov (USSR) (3 (15).04.1896–25.09.1986)*	"for their researches into the mechanism of chemical reactions"
1957	*Lord (Alexander R.) Todd (UK) (02.10.1907–10.01.1997)*	"for his work on nucleotides and nucleotide co-enzymes"
1958	*Frederick Sanger (UK) (b. 13.08.1918)*	"for his work on the structure of proteins, especially that of insulin"
1959	*Jaroslav Heyrovský (Czechoslovakia) (20.12.1890–27.03.1967)*	"for his discovery and development of the polarographic methods of analysis"
1960	*Willard Frank Libby (USA) (17.12.1908–08.09.1980)*	"for his method to use carbon-14 for age determination in archaeology, geology, geophysics, and other branches of science"
1961	*Melvin Calvin (USA) (07.04.1911–08.01.1997)*	"for his research on the carbon dioxide assimilation in plants"
1962	*Max Ferdinand Perutz (UK) (19.05.1914–06.02.2002) and John Cowdery Kendrew (UK) (24.03.1917–23.08.1997)*	"for their studies of the structures of globular proteins"

List of Nobel laureates in Chemistry (Continued)

1963	*Karl Ziegler (West Germany) (26.11.1898–11.08.1973) and Giulio Natta (Italy) (26.02.1903–02.05.1979)*	"for their discoveries in the field of the chemistry and technology of high polymers"
1964	*Dorothy Crowfoot Hodgkin (UK) (12.05.1910–29.07.1994)*	"for her determinations by X-ray techniques of the structures of important biochemical substances"
1965	*Robert Burns Woodward (USA) (10.04.1917–08.07.1979)*	"for his outstanding achievements in the art of organic synthesis"
1966	*Robert S. Mulliken (USA) (07.06.1896–31.10.1986)*	"for his fundamental work concerning chemical bonds and the electronic structure of molecules by the molecular orbital method"
1967	*½ – Manfred Eigen (West Germany) (b.09.05.1927)* *½ – Ronald George Wreyford Norrish (UK) (09.11.1897–07.06.1978) and George Porter (UK) (06.10.1920–31.08.2002)*	"for their studies of extremely fast chemical reactions, effected by disturbing the equilibrium by means of very short pulses of energy"
1968	*Lars Onsager (USA) (27.11.1903–05.10.1976)*	"for the discovery of the reciprocal relations bearing his name, which are fundamental for the thermodynamics of irreversible processes"
1969	*Derek H. R. Barton (UK) (08.08.1918–16.03.1998) and Odd Hassel (Norway) (17.05.1897–11.05.1981)*	"for their contributions to the development of the concept of conformation and its application in chemistry"
1970	*Luis F. Leloir (Argentine) (06.09.1906–02.12.1987)*	"for his discovery of sugar nucleotides and their role in the biosynthesis of carbohydrates"
1971	*Gerhard Herzberg (Canada) (25.12.1904–03.03.1999)*	"for his contributions to the knowledge of electronic structure and geometry of molecules, particularly free radicals"
1972	*½–Christian B.Anfinsen (USA) (26.03.1916–14.05.1995)*	"for his work on ribonuclease, especially concerning the connection between the amino acid sequence and the biologically active conformation"
	½ – Stanford Moore (USA) (04.09.1913–23.08.1982) and William H. Stein (USA) (25.06.1911–02.02.1980)	"for their contribution to the understanding of the connection between chemical structure and catalytic activity of the active centre of the ribonuclease molecule"
1973	*Ernst Otto Fischer (West Germany) (10.11.1918–23.07.2007) and Geoffrey Wilkinson (UK) (14.06.1921–26.09.1996)*	"for their pioneering work, performed independently, on the chemistry of the organometallic, so called sandwich compounds"
1974	*Paul J. Flory (USA) (19.06.1910–09.09.1985)*	"for his fundamental work, both theoretical and experimental, in the physical chemistry of macromolecules"
1975	*John Warcup Cornforth (Australia) (b. 07.09.1917)*	"for his work on the stereochemistry of enzyme-catalyzed reactions"
	Vladimir Prelog (Switzerland) (23.07.1906–07.01.1998)	"for his research into the stereochemistry of organic molecules and reactions"

1976	*William N. Lipscomb (USA) (b.09.12.1919)*	"for his studies on the structure of boranes illuminating problems of chemical bonding"
1977	*Ilya Prigogine (Belgium) (25.01.1917–28.05.2003)*	"for his contributions to non-equilibrium thermodynamics, particularly the theory of dissipative structures"
1978	*Peter D. Mitchell (UK) (29.09.1920–10.04.1992)*	"for his contribution to the understanding of biological energy transfer through the formulation of the chemiosmotic theory"
1979	*Herbert C. Brown (USA) (22.05.1912–1992 г.) and Georg Wittig (West Germany) (16.06.1897–26.08.1987)*	"for their development of the use of boron- and phosphorus-containing compounds, respectively, into important reagents in organic synthesis"
1980	*½ – Paul Berg (USA) (b. 30.06.1926)*	"for his fundamental studies of the biochemistry of nucleic acids, with particular regard to recombinant-DNA"
	½ – Walter Gilbert (USA) (b. 21.03.1932) and Frederick Sanger (UK) (b. 13.08.1980)	"for their contributions concerning the determination of base sequences in nucleic acids"
1981	*Kenichi Fukui (Japan) (b. 1918) and Roald Hoffmann (USA) (b. 18.07.1937)*	"for their theories, developed independently, concerning the course of chemical reactions"
1982	*Aaron Klug (UK) (b. 11.08.1926)*	"for his development of crystallographic electron microscopy and his structural elucidation of biologically important nucleic acid-protein complexes"
1983	*Henry Taube (USA) (1915–2005)*	"for his work on the mechanisms of electron transfer reactions, especially in metal complexes"
1984	*Robert Bruce Merrifield (USA) (15.07.1921–14.05.2006)*	"for his development of methodology for chemical synthesis on a solid matrix"
1985	*Herbert A. Hauptman (USA) (b. 1917) and Jerome Karle (USA) (b. 1918)*	"for their outstanding achievements in developing direct methods for the determination of crystal structures"
1986	*Dudley R.Herschbach (USA) (b. 1932), Yuan T. Lee (USA) (b. 1936) and John C. Polanyi (Canada) (b. 1929)*	"for their contributions concerning the dynamics of chemical elementary processes"
1987	*Donald J. Cram (USA) (b. 1919), Jean-Marie Lehn (France) (b. 30.09.1939) and Charles J. Pedersen (USA) (1904–1989)*	"for their development and use of molecules with structure-specific interactions of high selectivity"
1988	*Johann Deisenhofer (West Germany) (b. 1943), Robert Huber (West Germany) (b. 20.02.1937) and Hartmut Michel (West Germany) (b. 1948)*	"for their determination of the three-dimensional structure of a photosynthetic reaction centre"
1989	*Sidney Altman (USA) (b. 1939) and Thomas Cech (USA) (b. 08.12.1947)*	"for their discovery of catalytic properties of RNA"
1990	*Elias James Corey (USA) (b. 1928)*	"for his development of the theory and methodology of organic synthesis"
1991	*Richard R. Ernst (Switzerland) (b. 1933)*	"for his contributions to the development of the methodology of high resolution nuclear magnetic resonance (NMR) spectroscopy"
1992	*Rudolph A. Marcus (USA) (b. 21.07.1923)*	"for his contributions to the theory of electron transfer reactions in chemical systems"

List of Nobel laureates in Chemistry (Continued)

1993	*Kary B. Mullis (USA) (b. 1944)*	"for contributions to the developments of methods within DNA-based chemistry [...] for his invention of the polymerase chain reaction (PCR) method»
	Michael Smith (Canada) (1932–2000)	"for contributions to the developments of methods within DNA-based chemistry [...] for his fundamental contributions to the establishment of oligonucleotide-based, site-directed mutagenesis and its development for protein studies"
1994	*George A. Olah (USA) (b. 22.05.1927)*	"for his contribution to carbocation chemistry"
1995	*Paul J. Crutzen (Germany) (b. 1933), Mario J. Molina (USA) (b. 1943) and F. Sherwood Rowland (USA) (b. 1927)*	"for their work in atmospheric chemistry, particularly concerning the formation and decomposition of ozone"
1996	*Robert F. Curl Jr. (USA) (b. 1933), Sir Harold W. Kroto (USA) and Richard E. Smalley (USA) (06.06.1943–28.10.2005)*	"for their discovery of fullerenes"
1997	*½ – Paul D. Boyer (USA) (b. 1918) and John E. Walker (USA) (b. 1941)*	"for their elucidation of the enzymatic mechanism underlying the synthesis of adenosine triphosphate (ATP)"
	½ – Jens C. Skou (Denmark) (b. 1918)	"for the first discovery of an ion-transporting enzyme, Na^+, K^+-ATPase"
1998	*Walter Kohn (USA) (b. 09.03.1923)*	"for his development of the density-functional theory"
	John A. Pople (USA) (1925–2004)	"for his development of computational methods in quantum chemistry"
1999	*Ahmed Zewail (USA) (b. 26.02.1946)*	"for his studies of the transition states of chemical reactions using femtosecond spectroscopy"
2000	*Alan J. Heeger (USA) (b. 1936), Alan G. MacDiarmid (USA) (1927–2007) and Hideki Shirakawa (Japan) (b. 1936)*	"for their discovery and development of conductive polymers"
2001	*William S. Knowles (USA) (b. 1917), Ryōji Noyori (Japan) (b. 03.09.1938)*	"for their work on chirally catalysed hydrogenation reactions"
	K. Barry Sharpless (USA) (b. 1941)	"for his work on chirally catalysed oxidation reactions"
2002	*½ – John B. Fenn (USA) (b. 1917) and Koichi Tanaka (Japan) (b. 1959)*	"for the development of methods for identification and structure analyses of biological macromolecules [...] for their development of soft desorption ionisation methods for mass spectrometric analyses of biological macromolecules"
	½ – Kurt Wüthrich (Switzerland) (b. 04.10.1938)	"for the development of methods for identification and structure analyses of biological macromolecules [...] for his development of nuclear magnetic resonance spectroscopy for determining the three-dimensional structure of biological macromolecules in solution"
2003	*Peter Agre (USA) (b. 1949)*	"for discoveries concerning channels in cell membranes [...] for the discovery of water channels"

	Roderick MacKinnon (USA) (b. 1956)	"for discoveries concerning channels in cell membranes [...] for structural and mechanistic studies of ion channels"
2004	*Aaron Ciechanover (Israel) b. 1948), Avram Hershko (Israel) (b. 31.12.1937) and Irwin Rose (USA) (b. 1926)*	"for the discovery of ubiquitin-mediated protein degradation"
2005	*Robert H. Grubbs (USA) (b. 1942), Richard R. Schrock (USA) (b. 1945) and Yves Chauvin (France) (b. 1930)*	"for the development of the metathesis method in organic synthesis"
2006	*Roger D. Kornberg (USA) (b. 24.04.1947)*	"for his studies of the molecular basis of eukaryotic transcription"
2007	*Gerhard Ertl (Germany) (10.10.1936)*	"for his studies of chemical processes on solid surfaces"
2008	*Osamu Shimomura (Japan) (b. 27.08.1928), Martin Chalfie (USA) (b. 1947) and Roger Y. Tsien (USA) (b. 1952)*	"for the discovery and development of the green fluorescent protein, GFP"
2009	*Venkatraman Ramakrishnan (UK) (b. 1952), Thomas A. Steitz (USA) (b. 23.08.1940) and Ada E. Yonath (Israel) (b. 22.06.1939)*	"for studies of the structure and function of the ribosome"
2010	*Richard F. Heck (USA) (b. 15.08.1931), Akira Suzuki (Japan) (b. 12.09.1930) and Ei-ichi Negishi (Japan) (b. 14.07.1935)*	"for palladium-catalyzed cross couplings in organic synthesis"
2011	*Dan Shechtman (Israel) (b. 24.01.1941)*	"for the discovery of quasicrystals"

List of Nobel laureates in Physiology or Medicine

1901	*Emil Adolf von Behring (Germany) (15.03.1854–31.03.1917)*	"for his work on serum therapy, especially its application against diphtheria, by which he has opened a new road in the domain of medical science and thereby placed in the hands of the physician a victorious weapon against illness and deaths"
1902	*Sir Ronald Ross (India) (13.05.1857–16.09.1932)*	"for his work on malaria, by which he has shown how it enters the organism and thereby has laid the foundation for successful research on this disease and methods of combating it"
1903	*Niels Ryberg Finsen (Denmark) (15.12.1860–24.09.1904)*	"[for] his contribution to the treatment of diseases, especially lupus vulgaris, with concentrated light radiation, whereby he has opened a new avenue for medical science"
1904	*Ivan Petrovich Pavlov (Russia) (26.09.1849–27.02.1936)*	"in recognition of his work on the physiology of digestion, through which knowledge on vital aspects of the subject has been transformed and enlarged"
1905	*Robert Koch (Germany) (11.12.1843–27.05.1910)*	"for his investigations and discoveries in relation to tuberculosis"
1906	*Camillo Golgi (Italy) (07.07.1843–21.01.1926) and Santiago Ramón y Cajal (Spain) (01.05.1852–17.10.1934)*	"in recognition of their work on the structure of the nervous system"
1907	*Charles Louis Alphonse Laveran (France) (18.06.1845–18.05.1922)*	"in recognition of his work on the role played by protozoa in causing diseases"

List of Nobel laureates in Physiology or Medicine (Continued)

1908	*Ilya Ilyich Mechnikov (Russia) (03.05.1845–02.07.1916) and Paul Ehrlich (Germany) (14.03.1854–20.08.1915)*	"in recognition of their work on immunity"
1909	*Emil Theodor Kocher (Switzerland) (25.08.1841–27.07.1917)*	"for his work on the physiology, pathology and surgery of the thyroid gland"
1910	*Albrecht Kossel (Germany) (16.09.1853–05.07.1927)*	"in recognition of the contributions to our knowledge of cell chemistry made through his work on proteins, including the nucleic substances"
1911	*Allvar Gullstrand (Sweden) (05.06.1862–28.07.1930)*	"for his work on the dioptrics of the eye"
1912	*Alexis Carrel (France) (28.06.1873–05.11.1944)*	"[for] his work on vascular suture and the transplantation of blood vessels and organs"
1913	*Charles Richet (France) (25.08.1850–04.12.1935)*	"[for] his work on anaphylaxis"
1914	*Robert Bárány (Austria) (22.04.1876–08.04.1936) (awarded in 1915)*	"for his work on the physiology and pathology of the vestibular apparatus"
1915	*Not awarded*	
1916	*Not awarded*	
1917	*Not awarded*	
1918	*Not awarded*	
1919	*Jules Bordet (Belgium) (1870–1961) (awarded in 1920)*	"for his discoveries relating to immunity"
1920	*Schack August Steenberg Krogh (Denmark) (15.11.1874–13.09.1949)*	"for his discovery of the capillary motor regulating mechanism"
1921	*Not awarded*	
1922	*Archibald Vivian Hill (UK) (26.09.1886–03.06.1977)* *Otto Fritz Meyerhof (Germany) (12.04.1884–06.10.1951) (awarded in 1923)*	"for his discovery relating to the production of heat in the muscle" "for his discovery of the fixed relationship between the consumption of oxygen and the metabolism of lactic acid in the muscle"
1923	*Frederick Grant Banting (Canada) (14.11.1891–21.02.1941) and John James Rickard Macleod (Scotland) (06.09.1876–16.03.1935)*	"for the discovery of insulin"
1924	*Willem Einthoven (Netherlands) (21.05.1860–28.09.1927)*	"for the discovery of the mechanism of the electrocardiogram"
1925	*Not awarded*	
1926	*Johannes Andreas Grib Fibiger (Denmark) (23.04.1867–30.01.1928) (awarded in 1927)*	"for his discovery of the Spiroptera carcinoma"
1927	*Julius Wagner-Jauregg (Austria) (07.03.1857–27.09.1940)*	"for his discovery of the therapeutic value of malaria inoculation in the treatment of dementia paralytica"
1928	*Charles Jules Henri Nicolle (France) (21.09.1886–28.02.1936)*	"for his work on typhus"

1929	*Christiaan Eijkman (Netherlands) (11.08.1858–05.11.1930)* *Sir Frederick Gowland Hopkins (UK) (20.06.1861–16.05.1947)*	"for his discovery of the antineuritic vitamin" "for his discovery of the growth-stimulating vitamins"
1930	*Karl Landsteiner (Austria) (14.06.1868–26.06.1943)*	"for his discovery of human blood groups"
1931	*Otto Heinrich Warburg (Germany) (08.10.1883–01.08.1970)*	"for his discovery of the nature and mode of action of the respiratory enzyme"
1932	*Sir Charles Scott Sherrington (UK) (27.11.1857–04.03.1952) and Edgar Douglas Adrian (UK) (30.11.1889–04.08.1977)*	"for their discoveries regarding the functions of neurons"
1933	*Thomas Hunt Morgan (USA) (25.09.1866–04.12.1945)*	"for his discoveries concerning the role played by the chromosome in heredity"
1934	*George Hoyt Whipple (USA) (28.08.1878–01.02.1976), George Richards Minot (USA) (02.12.1885–25.02.1950) and William Parry Murphy (USA) (06.02.1892–09.10.1987)*	"for their discoveries concerning liver therapy in cases of anaemia"
1935	*Hans Spemann (Germany) (27.06.1869–09.09.1941)*	"for his discovery of the organizer effect in embryonic development"
1936	*Sir Henry Hallett Dale (UK) (09.06.1875–23.07.1968) and Otto Loewi (Germany) (03.06.1873–25.12.1961)*	"for their discoveries relating to chemical transmission of nerve impulses"
1937	*Albert Szent-Györgyi von Nagyrapolt (USA) (16.09.1893–22.10.1986)*	"for his discoveries in connection with the biological combustion processes, with special reference to vitamin C and the catalysis of fumaric acid"
1938	*Corneille Jean François Heymans (Belgium) (28.03.1892–18.07.1968) (awarded in 1939)*	"for the discovery of the role played by the sinus and aortic mechanisms in the regulation of respiration"
1939	*Gerhard Domagk (Germany) (30.10.1895–24.04.1964)*	"for the discovery of the antibacterial effects of prontosil"
1940	*Not awarded*	
1941	*Not awarded*	
1942	*Not awarded*	
1943	*Carl Peter Henrik Dam (Denmark) (21.02.1895–17.04.1976)* *Edward Adelbert Doisy (USA) (03.11.1893–23.10.1986)*	"for his discovery of vitamin K" "for his discovery of the chemical nature of vitamin K"
1944	*Joseph Erlanger (USA) (05.01.1874–05.12.1965) and Herbert Spencer Gasser (USA) (05.07.1888–11.05.1963)*	"for their discoveries relating to the highly differentiated functions of single nerve fibres"
1945	*Sir Alexander Fleming (UK) (06.08.1881–11.03.1955), Sir Ernst Boris Chain (UK) (19.06.1906–12.08.1979), Howard Walter Florey (UK) (24.09.1898–21.02.1968)*	"for the discovery of penicillin and its curative effect in various infectious diseases"
1946	*Hermann Joseph Muller (USA) (21.12.1890–05.04.1967)*	"for the discovery of the production of mutations by means of X-ray irradiation"

List of Nobel laureates in Physiology or Medicine (Continued)

1947	*½ – Carl Ferdinand Cori (USA) (05.12.1896–20.10.1984), Gerty Theresa Cori,née Radnitz (USA) (15.08.1896–26.10.1957)*	"for their discovery of the course of the catalytic conversion of glycogen"
	½ – Bernardo Alberto Houssay (Argentine) (10.04.1887–21.09.1971)	"for his discovery of the part played by the hormone of the anterior pituitary lobe in the metabolism of sugar"
1948	*Paul Hermann Müller (Switzerland) (12.01.1899–12.10.1965)*	"for his discovery of the high efficiency of DDT as a contact poison against several arthropods"
1949	*Walter Rudolf Hess (Switzerland) (17.03.1881–12.08.1973)*	"for his discovery of the functional organization of the interbrain as a coordinator of the activities of the internal organs"
	António Caetano Egas Moniz (Portugal) (29.11.1874–13.12.1955)	"for his discovery of the therapeutic value of leucotomy (lobotomy) in certain psychoses"
1950	*Edward Calvin Kendall (USA) (08.03.1886–04.05.1972), Tadeusz Reichstein (Switzerland) (20.07.1897–01.08.1996) and Philip Showalter Hench (USA) (28.02.1896–30.03.1965)*	"for their discoveries relating to the hormones of the adrenal cortex, their structure and biological effects"
1951	*Max Theiler (USA) (30.01.1899–11.08.1972)*	"for his discoveries concerning yellow fever and how to combat it"
1952	*Selman Abraham Waksman (USA) (02.07.1888–16.08.1973)*	"for his discovery of streptomycin, the first antibiotic effective against tuberculosis"
1953	*Sir Hans Adolf Krebs(UK) (25.08.1900–22.11.1981)*	"for his discovery of the citric acid cycle"
	Fritz Albert Lipmann (USA) (12.06.1899–24.07.1986)	"for his discovery of co-enzyme A and its importance for intermediary metabolism"
1954	*John Franklin Enders (USA) (10.02.1897–08.09.1985), Thomas Huckle Weller (USA) (b. 15.06.1915) and Frederick Chapman Robbins (USA) (25.08.1916–04.08.2003)*	"for their discovery of the ability of poliomyelitis viruses to grow in cultures of various types of tissue"
1955	*Axel Hugo Theodor Theorell (Sweden) (06.07.1903–15.08.1982)*	"for his discoveries concerning the nature and mode of action of oxidation enzymes"
1956	*André Frédéric Cournand (USA) (24.09.1895–19.02.1988), Werner Forssmann (West Germany) (29.08.1904–01.06.1979) and Dickinson W. Richards (USA) (30.10.1895–23.02.1973)*	"or their discoveries concerning heart catheterization and pathological changes in the circulatory system"
1957	*Daniel Bovet (Italy) (23.03.1907–08.04.1992)*	"for his discoveries relating to synthetic compounds that inhibit the action of certain body substances, and especially their action on the vascular system and the skeletal muscles"
1958	*½ – George Wells Beadle (USA) (22.10.1903–09.06.1989) and Edward Lawrie Tatum (USA) (14.12.1909–05.11.1975)*	"for their discovery that genes act by regulating definite chemical events"
	½ – Joshua Lederberg (USA) (b. 23.05.1925)	"for his discoveries concerning genetic recombination and the organization of the genetic material of bacteria"

1959	*Severo Ochoa (USA) (24.09.1905–01.11.1993) and Arthur Kornberg (USA) (03.03.1918–26.10.2007)*	"for their discovery of the mechanisms in the biological synthesis of ribonucleic acid and deoxyribonucleic acid"
1960	*Sir Frank Macfarlane Burnet (Australia) (03.09.1899–31.08.1985) and Sir Peter Brian Medawar (UK) (28.02.1915–02.10.1987)*	"for discovery of acquired immunological tolerance"
1961	*Georg von Békésy (USA) (03.06.1899–13.06.1972)*	"for his discoveries of the physical mechanism of stimulation within the cochlea"
1962	*Francis Harry Compton Crick (UK) (b. 08.06.1916), James Dewey Watson (USA) (b. 06.04.1928) and Maurice Hugh Frederick Wilkins (USA) (15.12.1916–05.10.2004)*	"for their discoveries concerning the molecular structure of nucleic acids and its significance for information transfer in living material"
1963	*Sir John Carew Eccles (Australia) (27.01.1903–02.05.1997), Sir Alan Lloyd Hodgkin (UK) (05.02.1914–20.12.1998) and Sir Andrew Fielding Huxley (UK) (b. 22.11.1917)*	"for their discoveries concerning the ionic mechanisms involved in excitation and inhibition in the peripheral and central portions of the nerve cell membrane"
1964	*Konrad Bloch (West Germany, USA) (21.01.1912–15.10.2000) and Feodor Lynen (West Germany) (06.04.1911–06.08.1979)*	"or their discoveries concerning the mechanism and regulation of the cholesterol and fatty acid metabolism"
1965	*François Jacob (France) (b.17.06.1920), André Lwoff (France) (08.05.1902–30.09.1994) and Jacques Monod (France) (09.02.1910–31.05.1976)*	"for their discoveries concerning genetic control of enzyme and virus synthesis"
1966	*Peyton Rous (USA) (05.10.1879–16.02.1970)* *Charles Brenton Huggins (USA) (22.09.1901–12.01.1997)*	"for his discovery of tumour-inducing viruses" "for his discoveries concerning hormonal treatment of prostatic cancer"
1967	*Ragnar Granit (Sweden) (30.10.1900–12.03.1991), Haldan Keffer Hartline (USA) (22.12.1903–17.03.1983) and George Wald (USA) (18.11.1906–12.04.1997)*	"for their discoveries concerning the primary physiological and chemical visual processes in the eye"
1968	*Robert W. Holley (USA) (28.01.1922–11.02.1993), Har Gobind Khorana (USA) (b. 09.01.1922) and Marshall W. Nirenberg (USA) (b. 10.04.1927)*	"for their interpretation of the genetic code and its function in protein synthesis"
1969	*Max Delbrück (USA) (04.09.1906–09.03.1981), Alfred D. Hershey (USA) (04.12.1908–22.05.1997) and Salvador E. Luria (USA) (13.08.1912–06.02.1991)*	"for their discoveries concerning the replication mechanism and the genetic structure of viruses"
1970	*Sir Bernard Katz (UK) (26.03.1911–20.04.2003), Ulf von Euler (Sweden) (07.02.1905–09.03.1983) and Julius Axelrod (USA) (30.05.1912–29.12.2004)*	"for their discoveries concerning the humoral transmittors in the nerve terminals and the mechanism for their storage, release and inactivation"
1971	*Earl W. Sutherland,Jr. (USA) (19.11.1915–09.03.1974)*	"for his discoveries concerning the mechanisms of the action of hormones"

List of Nobel laureates in Physiology or Medicine (Continued)

1972	*Gerald M. Edelman (USA) (b. 01.07.1929) and Rodney R. Porter (UK) (08.10.1917–07.09.1985)*	"for their discoveries concerning the chemical structure of antibodies"
1973	*Karl von Frisch (Austria) (20.11.1886–12.06.1982), Konrad Lorenz (Austria) (07.11.1903–27.02.1989) and Nikolaas Tinbergen (Niderlands) (15.04.1907–21.12.1988)*	"for their discoveries concerning organization and elicitation of individual and social behaviour patterns"
1974	*Albert Claude (Belgium, USA) (23.08.1899–22.05.1983), Christian de Duve (Belgium) (b. 02.10.1917) and George E. Palade (USA) (19.11.1912–08.10.2008)*	"for their discoveries concerning the structural and functional organization of the cell"
1975	*David Baltimore (USA) (b. 07.03.1932), Renato Dulbecco (Italy) (b. 22.02.1914) and Howard Martin Temin (USA) (10.12.1934–09.02.1994)*	"for their discoveries concerning the interaction between tumour viruses and the genetic material of the cell"
1976	*Baruch S. Blumberg (USA) (b. 28.07.1925) and D. Carleton Gajdusek (USA) (b. 09.09.1923)*	"for their discoveries concerning new mechanisms for the origin and dissemination of infectious diseases"
1977	*½ – Roger Guillemin (France) (b. 11.01.1924) and Andrew V. Schally (Polonia) (b. 30.11.1926)*	"for their discoveries concerning the peptide hormone production of the brain"
	½ – Rosalyn Yalow (USA) (b. 19.07.1921)	"for the development of radioimmunoassays of peptide hormones"
1978	*Werner Arber (Switzerland) (b. 03.06.1929), Daniel Nathans (USA) (30.10.1928–16.11.1999) and Hamilton O. Smith (USA) (b. 23.08.1931)*	"for the discovery of restriction enzymes and their application to problems of molecular genetics"
1979	*Allan M. Cormack (USA) (23.02.1924–07.05.1998) and Sir Godfrey N. Hounsfield (UK) (28.08.1919–12.08.2004)*	"for the development of computer assisted tomography"
1980	*Baruj Benacerraf (USA) (b. 29.10.1920), Jean Dausset (USA) (b. 19.10.1916) and George D. Snell (USA) (19.12.1903–06.06.1996)*	"for their discoveries concerning genetically determined structures on the cell surface that regulate immunological reactions"
1981	*½ – Roger W. Sperry (USA) (20.08. 1913–17.04.1994)*	"for his discoveries concerning the functional specialization of the cerebral hemispheres"
	½ – David H. Hubel (USA) (b. 27.02.1926) and Torsten N. Wiesel (Sweden) (b. 03.06.1924)	"for their discoveries concerning information processing in the visual system"
1982	*Sune K. Bergström (Sweden) (10.01.1916–15.08.2004), Bengt I. Samuelsson (Sweden) (b. 21.05.1934) and Sir John R. Vane (UK) (29.03.1927–19.11.2004)*	"for their discoveries concerning prostaglandins and related biologically active substances"
1983	*Barbara McClintock (USA) (16.06.1902–02.09.1992)*	"for her discovery of mobile genetic elements"

1984	*Niels K. Jerne (Denmark) (23.12.1911–07.10.1994), Georges J.F. Köhler (West Germany) (17.04.1946–01.03.1995) and César Milstein (Argentine) (08.10.1927–24.03.2002)*	"for theories concerning the specificity in development and control of the immune system and the discovery of the principle for production of monoclonal antibodies"
1985	*Michael S. Brown (USA) (b. 13.04.1941) and Joseph L. Goldstein (USA) (b. 18.04.1940)*	"for their discoveries concerning the regulation of cholesterol metabolism"
1986	*Stanley Cohen (USA) (b. 17.11.1922) and Rita Levi-Montalcini (USA)*	"for their discoveries of growth factors"
1987	*Susumu Tonegawa (Japan) (b. 06.09.1939)*	"for his discovery of the genetic principle for generation of antibody diversity"
1988	*Sir James W. Black (Scotland) (b. 14.07.1924), Gertrude B. Elion (USA) (23.01.1918–21.02.1999) and George H. Hitchings (USA) (18.04.1905–27.02.1998)*	"for their discoveries of important principles for drug treatment"
1989	*J. Michael Bishop (USA) (b. 22.02.1936) and Harold E. Varmus (USA) (b. 18.12.1939)*	"for their discovery of the cellular origin of retroviral oncogenes"
1990	*Joseph E. Murray (USA) (b. 01.04.1919) and E. Donnall Thomas (USA) (b. 15.03.1920)*	"for their discoveries concerning organ and cell transplantation in the treatment of human disease"
1991	*Erwin Neher (Germany) (b. 20.03.1944) and Bert Sakmann (Germany) (b. 12.06.1942)*	"for their discoveries concerning the function of single ion channels in cells"
1992	*Edmond H. Fischer (USA) (b. 06.04.1920) and Edwin G. Krebs (USA) (b. 06.06.1918)*	"for their discoveries concerning reversible protein phosphorylation as a biological regulatory mechanism"
1993	*Sir Richard J. Roberts (UK) (b. 06.09.1943) and Phillip A. Sharp (USA) (b. 06.06.1944)*	"for their discoveries of split genes"
1994	*Alfred G. Gilman (USA) (b. 01.07.1941) and Martin Rodbell (USA) (01.12.1925–07.12.1998)*	"for their discovery of G-proteins and the role of these proteins in signal transduction in cells"
1995	*Edward B. Lewis (USA) (20.05.1918–21.07.2004), Christiane Nüsslein-Volhard (Germany) (b. 20.10.1942) and Eric F. Wieschaus (USA) (b. 08.06.1947)*	"for their discoveries concerning the genetic control of early embryonic development"
1996	*Peter C. Doherty (Australia) (b. 15.10.1940) and Rolf M. Zinkernagel (Switzerland) (b. 06.01.1944)*	"for their discoveries concerning the specificity of the cell mediated immune defence"
1997	*Stanley B. Prusiner (USA) (b. 28.05.1942)*	"for his discovery of Prions—a new biological principle of infection"
1998	*Robert F. Furchgott (USA) (b. 04.06.1916), Louis J. Ignarro (USA) (b. 31.05.1941) and Ferid Murad (USA) (b. 14.09.1936)*	"for their discoveries concerning nitric oxide as a signalling molecule in the cardiovascular system"
1999	*Günter Blobel (Germany) (b. 21.05.1936)*	"for the discovery that proteins have intrinsic signals that govern their transport and localization in the cell"
2000	*Arvid Carlsson (Sweden) (b. 25.01.1923), Paul Greengard (USA) (b. 11.12.1925) and Eric R. Kandel (USA) (b. 07.11.1929)*	"for their discoveries concerning signal transduction in the nervous system"

List of Nobel laureates in Physiology or Medicine (Continued)

2001	*Leland H. Hartwell (USA) (b. 30.10.1939), Sir Tim Hunt (UK) (b. 19.02.1943) and Sir Paul M. Nurse (UK) (b. 25.01.1949)*	"for their discoveries of key regulators of the cell cycle"
2002	*Sydney Brenner (IUAR) (b. 13.01.1927), H. Robert Horvitz (USA) (b. 08.05.1947) and Sir John E. Sulston (UK) (b. 27.03.1942)*	"for their discoveries concerning 'genetic regulation of organ development and programmed cell death"
2003	*Paul Lauterbur (USA) (06.05.1929–27.03.2007) and Sir Peter Mansfield (UK) (b. 09.10.1933)*	"for their discoveries concerning magnetic resonance imaging"
2004	*Richard Axel (USA) (b. 02.07.1946) and Linda B. Buck (USA) (b. 29.01.1947)*	"for their discoveries of odorant receptors and the organization of the olfactory system"
2005	*Barry J. Marshall (Australia) (b. 30.09.1951) and J. Robin Warren (Australia) (b. 11.06.1937)*	"for their discovery of the bacterium Helicobacter pylori and its role in gastritis and peptic ulcer disease"
2006	*Andrew Z. Fire (USA) (b. 27.04.1959) and Craig C. Mello (USA) (b. 18.10.1960)*	"for their discovery of RNA interference - gene silencing by double-stranded RNA"
2007	*Mario R. Capecchi (USA) (b. 06.10.1937), Sir Martin J. Evans (UK) (b. 01.01.1941) and Oliver Smithies (UK) (b. 23.07.1925)*	"for their discoveries of principles for introducing specific gene modifications in mice by the use of embryonic stem cells"
2008	*Harald zur Hausen (Germany) (b. 11.0.1936)* *Françoise Barré-Sinoussi (France) (b. 30.07.1947) and Luc Montagnier (France) (b. 18.08.1932)*	"for his discovery of human papilloma viruses causing cervical cancer" "for their discovery of human immunodeficiency virus"
2009	*Elizabeth H.Blackburn (USA–Australia) (b. 26.11.1948), Jack W. Szostak (USA) (09.11.1952) and Carol W. Greider (USA) (15.04.1961)*	"for the discovery of how chromosomes are protected by telomeres and the enzyme telomerase"
2010	*Sir Robert G. Edwards (UK) (b. 1925)*	"for the development of in vitro fertilization"
2011	*Bruce A. Beutler (USA) (b. 29.12.1957) and Jules A. Hoffmann (France) (b. 02.08.1941)* *Ralph M. Steinman (Canada–USA) (14.01.1943–30.09.2011)* (awarded posthumously)	"for their discoveries concerning the activation of innate immunity" "for his discovery of the dendritic cell and its role in adaptive immunity"

INDEX

A

B

C

D

E

F

I

J

K

L

M

N

O

P

Q

T

U

V

W

X

Y

Z